Conference Proceedings of the Society for Experimental Mechanics Series

Series Editor

Kristin B. Zimmerman
Society for Experimental Mechanics, Inc.,
Bethel, USA

The Conference Proceedings of the Society for Experimental Mechanics Series presents early findings and case studies from a wide range of fundamental and applied work across the broad range of fields that comprise Experimental Mechanics. Series volumes follow the principle tracks or focus topics featured in each of the Society's two annual conferences: IMAC, A Conference and Exposition on Structural Dynamics, and the Society's Annual Conference & Exposition and will address critical areas of interest to researchers and design engineers working in all areas of Structural Dynamics, Solid Mechanics and Materials Research.

Garrett Pataky · John Kolinski · Thomas Berfield · Nicholas Bachus ·
Rosa De Finis · Suhasini Gururaja
Editors

Fracture, Fatigue, Failure, Damage Evolution and Thermomechanics & Infrared Imaging, Vol. 4

Proceedings of the 2025 Annual Conference
on Experimental and Applied Mechanics

River Publishers

Published, sold and distributed by:

River Publishers
Broagervej 10
9260 Gistrup
Denmark

www.riverpublishers.com

ISBN 97887-438-0830-5 (Hardback)

ISBN 97887-438-0835-0 (eBook)

https://doi.org/10.13052/97887-438-0830-5

Conference Proceedings of the Society for Experimental Mechanics

An imprint of River Publishers

Preface

Fracture, Fatigue, Failure, Damage Evolution and *Thermomechanics & Infrared Imaging* represents one of five volumes of technical papers presented at the SEM 2025 SEM Annual Conference & Exposition on Experimental and Applied Mechanics organized by the Society for Experimental Mechanics and held in Milwaukee, WI, June 2-5, 2025. The complete Proceedings also includes volumes on: *Dynamic Behavior of Materials; Challenges in Mechanics of Time-Dependent Materials; Advancement of Optical Methods & Digital Image Correlation in Experimental Mechanics; Mechanics of Biological Systems and Materials and the Mechanics of Composite, Hybrid & Multifunctional Materials; and Mechanics of Additive & Advanced Manufacturing, Inverse Methods and Machine Learning.*

Each collection presents early findings from experimental and computational investigations on an important area within Experimental Mechanics, Fracture and Fatigue being one of these areas.

Fatigue and fracture are two of the most critical considerations in engineering design. Understanding and characterizing fatigue and fracture has remained as one of the primary focus areas of experimental mechanics for several decades. Advances in experimental techniques, such as digital image correlation, acoustic emissions, and electron microscopy, have allowed for deeper study of phenomena related to fatigue and fracture. This volume contains the results of investigations of several aspects of fatigue and fracture such as microstructural effects, the behavior of interfaces, the behavior of different and/or complex materials such as composites, and environmental and loading effects. The collection of experimental mechanics research included here represents another step toward solving the long-term challenges associated with fatigue and fracture.

In recent years the applications of infrared imaging techniques to the mechanics of materials and structures has grown considerably. The expansion is marked by the increased spatial and temporal resolution of the infrared detectors, faster processing times, much greater temperature resolution and specific image processing. The improved sensitivity and more reliable temperature calibrations of the devices have meant that more accurate data can be obtained than were previously available.

Editors:
Garrett Pataky-*Clemson University, USA; John Kolinski-EPFL, Switzerland;* Thomas Berfield, *University of Louisville, KY, USA*; Nicholas Bachus, University of California Davis, CA, USA; Rosa De Finis, *University of Salento, Italy*; Suhasini Gururaja, *Auburn University, AL, USA*

Contents

Chapter 1
Rapid Fatigue Characterisation of AlSi10Mg-AM using Energy-based approaches

Rosa De Finis, Davide Palumbo, and Umberto Galietti

Abstract The advent of additive manufacturing has transformed the manner in which structural and non-structural materials and components are designed and manufactured. This technology offers the flexibility of manufacturing but also presents a number of challenges concerning the performance of the material itself.

One such challenge is the lack of data regarding the structural aspects of the material, including fatigue resistance. This information is crucial for understanding the behaviour of a component under operating loads.

A valuable support in such activities aimed at shedding light on dynamic behaviour is provided by the techniques provided by the experimental mechanics. In particular, Infrared thermography is a valuable tool for detecting anomalies in the thermal map of a material or component subjected to cyclic loading. This is achieved using certain indices linked to reversible and irreversible physical processes that characterise the nature of heat sources. Furthermore, conducting rapid fatigue tests allows for the study of material behaviour through the examination of material self-heating.

The aim of this study is to present the challenging aspects of studying the fatigue behaviour of aluminium alloys produced by Additive Manufacturing. This is regarding both the behaviour of aluminium from a thermal point of view (it is a highly diffusive material, so temperature variations are very low) and the behaviour from a mechanical point of view (the fatigue limit of some alloys does not present the characteristic "knee" point). The goal is to demonstrate how thermography can support structural evaluations on this material.

Introduction

Additive manufacturing is a new material production technique that combines several technological advantages such as the ability to build complex-shaped components in a relatively short time using a relatively circular process and, in some cases, no material rework is required [1]. However, there are still many challenges to face like induced anisotropy, residual stresses, surface finish that affect structural integrity and mechanical material properties. Indeed, the characterization of heterogeneous or anisotropic microstructures, along with the presence of residual stresses and induced defects, and the dependence of material properties on building direction and orientation, requires the development of ad-hoc testing and analysis procedures. Specifically, the whole comprehension of the fatigue behavior in these materials remain not fully achieved. This push the research toward new methodologies for studying the fatigue behavior [1–2].

The experimental mechanics provides invaluable support for both structural integrity inspections and material characterization [3–9]. In particular, in the past different thermography-based techniques were developed to carry out the rapid characterization of materials and the early damage detection. These techniques present the advantage of assessing the damage information in relatively short time. Such techniques are mainly adopted during rapid fatigue tests that exploit the self-heating effect of the material. In these tests by fatigue loading the material first at very low and then at very high load levels until failure, the transition between damage regimes can be studied [3–5]. In fact, if a thermal imaging camera monitors the test, it is possible to derive a range of information related to the energy involved in dissipative processes that serve

Rosa De Finis (✉)
Dipartimento di Ingegneria dell'Innovazione, University of Salento, Via per monteroni, 73100 Lecce, Italy
e-mail: rosa.definis@unisalento.it

Davide Palumbo · Umberto Galietti
Dipartimento di Meccanica, Matematica e Management, Politecnico di Bari, Via Orabona, 5-70125 Bari, Italy

Garrett Pataky et al. (eds.), *Fracture, Fatigue, Failure, Damage Evolution and Thermomechanics & Infrared Imaging, Vol. 4*,
Conference Proceedings of the Society for Experimental Mechanics Series, https://doi.org/10.13052/9788743808305-1

as a sentinel for early and local detection of damage. This has enabled the development of energy-based approaches and methodologies [10–18].

Energy-based methodologies [6–9], founded upon the evaluation of quantifiable variables such as temperature and strain, have exhibited their efficacy in the identification of damage, even in the presence of small surface defects, and in the separation of inelastic from anelastic material behavior [10, 12, 17]. More in detail, in recent years, the estimation of energy dissipated during fatigue processes through the assessment of temperature fluctuations (second amplitude harmonics, or SAH [7–9]) has demonstrated considerable potential for the precise analysis of material behavior through the assessment of a parameter that is resistant to the influence of numerous disturbing noise factors.

This paper presents preliminary results of fatigue characterization of the AlSi10Mg-AM alloy produced by selective laser melting, using rapid methods. The SAH and the area under hysteresis loop are used as damage indices and to study the material behavior.

The results will also be analyzed in terms of material specificity: SAH will represent not only dissipative but also thermoelastic effects [11, 18]. This significantly complicates the assessment of material behavior.

Moreover, an ad-hoc procedure for the evaluation of material fatigue strength taking into account the specific thermal behavior of the material is also provided.

Materials and Methods

The present study dealt with AlSi10Mg alloy produced by selective laser melting additive manufacturing process [2]. The samples were built in vertically direction and heat-treated (age-hardening for 6 hours-T6). Moreover, the surface of the samples was matt black coated to enhance and uniform the surface emissivity.

All fatigue tests were performed on a servo-hydraulic MTS 647 loading machine with a 100kN load cell. The samples were addressed to a stepwise loading procedure in order to study the self-heating material behaviour [3–10]. It consists of a sequence of load blocks, each with the same duration of load cycles and fixed values for load frequency and stress ratio. The sequence of blocks is defined by incrementally increasing the applied stress level. Three acquisitions were carried out at each load step at 1000 and 15000 cycles respectively at a loading frequency of 15hz which is sufficiently high to avoid loss of adiabatic conditions. To acquire thermal data the Flir X6540SC Infrared camera was adopted. In particular, this camera has a cooled detector with 640×512 pixels matrix and a NETD lower than 25 mK. The adopted frame rate to obtain IR sequences was 200Hz. An extensometer MTS 634 (clip-on type with 25 mm gage length) acquired the strain data at 500 frame per second.

The setup and equipment are reported in Fig. 1.

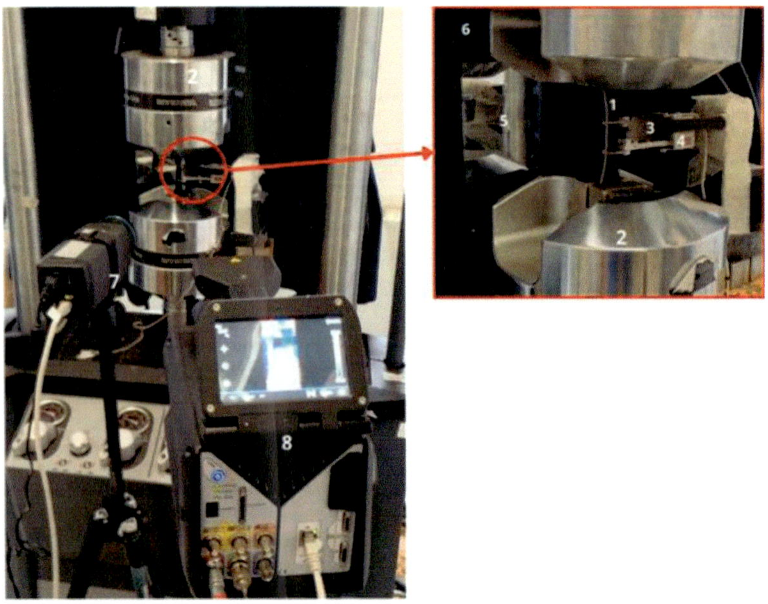

Fig. 1 Experimental setup (1:sample, 2:clamps, 3:dummy specimen;4: extensometer, 5:mirror, 6: cover sheet; 7:microbolometer camera; 8:cooled camera FLIR X6540SC).

Referring to the signal processing a signal reconstruction algorithm based on the least squares method [7] is adopted. The algorithms are implemented in the software IRTA®. The mathematical model describing the temeprature evolution and fluctuation is:

$$T(t) = T_0 + a \times t + T_1 \times \sin\left[2\pi f_L\, t + \varphi_1\right] + T_2 \times \sin\left[4\pi f_L\, t + \varphi_2\right] \tag{1}$$

where T_0 is the mean value, a is the mean rate, T_1 and T_2 the two amplitudes, and φ_1 and φ_2 the two phases and f_L is the loading frequency.

The processing was performed pixel by pixel but the identified indicator for the damage was the 95° percentile value in a region of interest (ROI) coinciding with the gage volume of the sample, was considered to represent the T_2 field.

The T_2 maps were addressed to a double-dimension filter using a gaussian kernel in order to filter out the noise.

The strain data from extensometer were used to study the hysteresis loop behaviour and in general the energy involved in fatigue processes.

Results

Fig. 2 shows the maps of assessed second amplitude harmonics (SAH) referring to the gage volume. The analysis of the SAH maps reveals an increase in signal from the first load levels to the last for all specimens. The damage appears to be widespread, affecting the entire useful section. However, it is only in the final loading steps that circumscribed areas with a markedly higher signal can be identified, which are indicative of the damage sites.

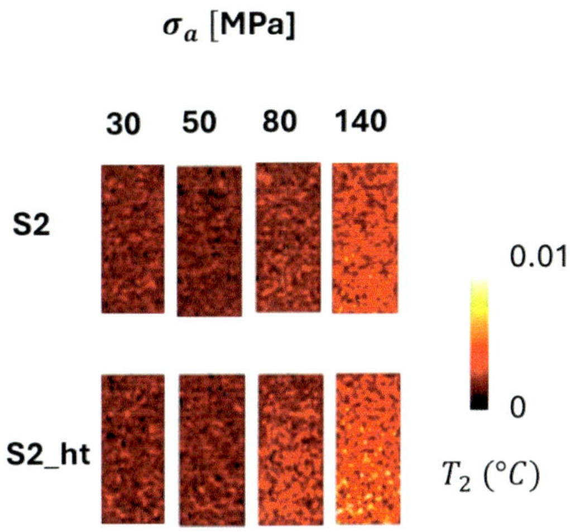

Fig. 2 Maps of SAH for samples as provided (right) and heat-treated (left).

Fig. 3 shows two exemplary charts with the hysteresis loops related to a sample in 'as provided' condition and a sample after heat treatment. The curves show that in both conditions the material behaviour is characterized by limited plastic

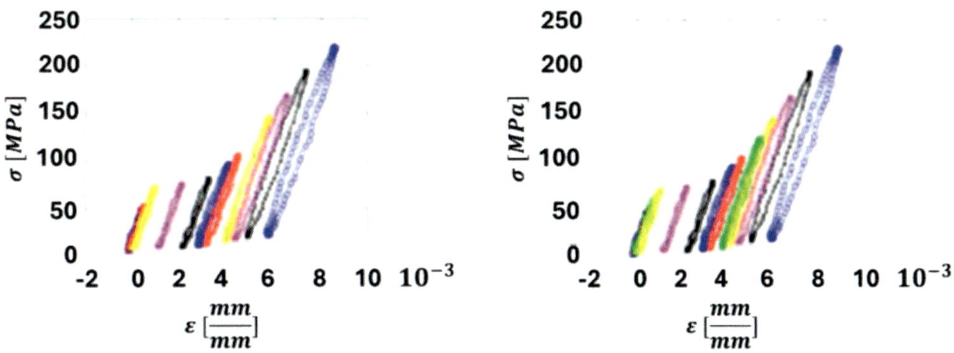

Fig. 3 Hysteresis loops curves for sample 3, as provided (right) and heat treated (left).

deformation and general brittle behaviour. The dissipation increases markedly only in the last loading levels where the hysteresis loops are wider than those at the beginning of the test.

Fig. 4 shows the curves of SAH (Fig. 4a) and the area under the hysteresis loop-Wp (Fig. 4b) for the specimens built in the vertical direction.

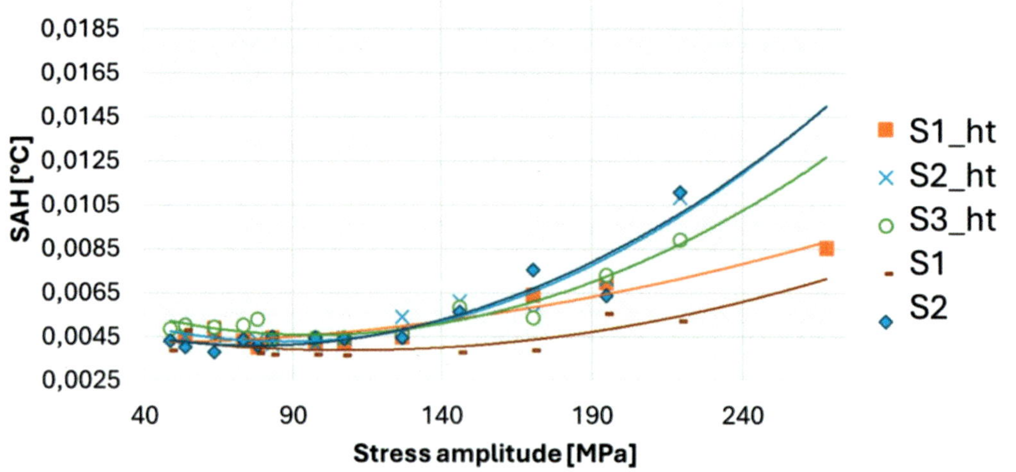

(a)

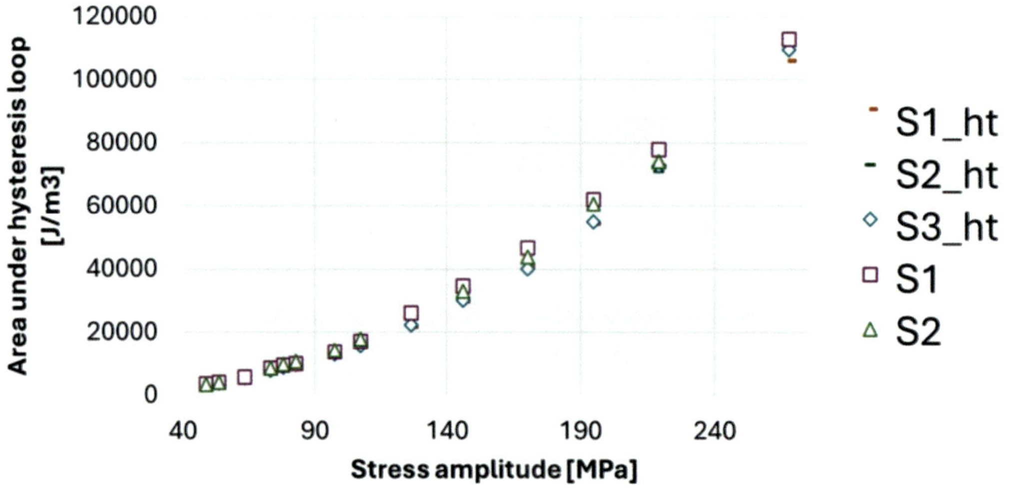

(b)

Fig. 4 Curves for samples realized in z-direction with and without heat treatment (ht=heat treatment), (a) SAH , (b) area under hysteresis loop (Wp).

As can be seen from Fig. 4a, there is no real cut-off point in the thermal behaviour of the material as would be expected from the reference parameter (Fig. 4b). This may be due to the superposition of reversible and irreversible effects in the thermal behaviour of the material. It should also be noted that aluminum does not have a knee point [16], but rather it is necessary to refer to fatigue strength at a given number of cycles. This will require future testing to better understand how to separate the two reversible and irreversible thermal contributions, as well as ad hoc processing procedures.

To estimate the material fatigue strength, a new procedure was identified to be applied to each specimen to select the stress level at which the series trend deviates significantly from the initial trend representing undamaged material.

The procedure is summarized as follows:

- Starting from the SAH values of the last 3 stress levels, consider the (stress amplitudes, SAH) pairs and calculate the R^2 coefficient;

- Adding another pair of data (stress amplitudes, SAH) and calculate the R^2 coefficient;
- Continuing until the value of R^2 differs by more than 5% from the previous values.

The methodology has been applied to SAH data and produced an estimation of the fatigue limit of 97±10 MPa without much difference between heat treated specimens and 'as provided' specimens.

Conclusions

In this paper, a comparison between SAH and hysteresis loop area has been presented to study the fatigue behaviour of AlSi10Mg -AM alloy subjected to stepwise testing.

The thermal method proves to be a comprehensive technique, allowing not only the detection of damage from its onset, but also the study of its evolution. All this has the advantage of consistently reducing the duration of the experimental campaign.

An additional advantage is the ability to better understand any differences between heat-treated and 'as provided' material, which is relevant for the purpose of providing guidelines for the technological process.

From the obtained results, there is in fact no difference between the fatigue strength of the material in the two configurations, but the data are certainly less noisy after heat treatment.

References

1. Liu, H., Yu, H., Guo, C., Chen, X., Zhong, S., Zhou, L., Osman, A., & Lu, J. (2023). Review on Fatigue of Additive Manufactured Metallic Alloys: Microstructure, Performance, Enhancement, and Assessment Methods.
2. Beretta, S., Gargourimotlagh, M., Foletti, S., du Plessis, A., & Riccio, M. (2020). Fatigue strength assessment of as built AlSi10Mg manufactured by SLM with different build orientations.
3. M.P. Luong, Fatigue limit evaluation of metals using an infrared thermographic technique, Mechanics of Materials, 28 (1998) 155-163.
4. G. Fargione, A. Geraci, G. La Rosa, A. Risitano, Rapid determination of the fatigue curve by the thermographic method, International Journal of Fatigue, 24 (2002) 11-19.
5. T. Boulanger, A. Chrysochoos, C. Mabru, A. Galtier, Calorimetric analysis of dissipative and thermoelastic effects associated with the fatigue behavior of steels, International Journal of Fatigue, 26 (2004) 221-229.
6. F. Maquin, F. Pierron, Heat dissipation measurements in low stress cyclic loading of metallic materials: From internal friction to microplasticity, Mechanics of Materials, 41 (2009) 928-942.
7. R. De Finis, D. Palumbo, U. Galietti, A multianalysis thermography-based approach for fatigue and damage investigations of ASTM A182 F6NM steel at two stress ratios, Fatigue & Fracture of Engineering Materials & Structures, 42 (2019) 267-283.
8. G. Meneghetti, M. Ricotta, Estimating the intrinsic dissipation using the second harmonic of the temperature signal in tension-compression fatigue. Part II: Experiments, Fatigue & Fracture of Engineering Materials & Structures, 44 (2021) 2153-2167.
9. R. De Finis, D. Palumbo, U. Galietti, On the relationship between mechanical energy rate and heat dissipated rate during fatigue for a C45 steel depending on stress ratio, Fatigue & Fracture of Engineering Materials & Structures, 44 (2021) 2781-2799.
10. Doudard C, Calloch S, Cugy P, Galtier A, Hild F. A probabilistic two-scale model for high-cycle fatigue life predictions. *Fatigue & Fracture of Engineering Materials & Structures.* 2005;28(3):279-288.
11. Wong AK, Sparrow JG, Dunn SA. On the revised theory of the thermoelastic effect. *Journal of Physics and Chemistry of Solids.* 1988;49(4):395-400.
12. Sakagami T, Kubo S, Tamura E, Nishimura T. Identification of plastic-zone based on double frequency lock-in thermographic temperature measurement. 2013.
13. Meneghetti G, Ricotta M. Evaluating the heat energy dissipated in a small volume surrounding the tip of a fatigue crack. *International Journal of Fatigue.* 2016;92:605-615.
14. Meneghetti G, Ricotta M. The heat energy dissipated in the material structural volume to correlate the fatigue crack growth rate in stainless steel specimens. *International Journal of Fatigue.* 2018;115:107-119.
15. Liakat M, Khonsari MM. On the anelasticity and fatigue fracture entropy in high-cycle metal fatigue. *Materials & Design.* 2015;82:18-27.
16. De Finis R, Palumbo D, Serio LM, De Filippis LAC, Galietti U. Correlation between thermal behaviour of AA5754-H111 during fatigue loading and fatigue strength at fixed number of cycles. *Materials.* 2018;11(5):719.
17. Lepitre P, Calloch S, Dhondt M, Surand M, Doudard C. Identification of the damage scenarios under cyclic loading of a coated 300M steel by infrared thermography measurements. *Physical Sciences Forum.* 2022;4(1):30.
18. Di Carolo, F.,De Finis, R., Palumbo, D., Galietti, U. A thermoelastic stress analysis general model: Study of the influence of biaxial residual stress on aluminium and titanium, Metals, 2019, 9(6), 671.

Chapter 2
Effect of Induced Porosity on Fatigue Performance of Additively Manufactured Short-Fiber Thermoplastics Using Infrared Thermography

Harsh Gandhi, Pharindra Pathak, and Suhasini Gururaja

Abstract This study explores the influence of induced porosity on the fatigue behavior of additively manufactured short-fiber thermoplastics (AM-SFTs), employing infrared thermography (IRT) as an accelerated characterization method. IRT has emerged as a validated technique for drastically reducing the time required to estimate fatigue limits, offering a more efficient alternative to the conventional stress-life (S-N) method. In a previous study, acquiring the S-N curve for 20% short carbon fiber-reinforced Acrylonitrile Butadiene Styrene (CF/ABS) panels required two weeks of testing with 18 specimens. IRT testing achieved comparable fatigue limit estimates using only three specimens over five hours. This underscores the effectiveness of IRT in providing rapid fatigue assessments [1]. In this research, specimens composed of 80% ABS and 20% chopped carbon fibers were produced via fused deposition modeling (FDM) and subjected to fatigue loading. The study considered 16-layer AM-SFT specimens with artificially induced discrete porosity - 6 mm diameter defects located on the middle layer of the specimen. Additionally, control specimens with no pores have also been tested. A test matrix of 15 specimens consistent with ASTM D638 Type I specifications, with five replicates per configuration, has been tested. Each specimen was tested under both static and cyclic loading conditions. IRT facilitated real-time monitoring of surface temperature changes during fatigue testing, a precursor to fatigue damage initiation. This investigation aims to establish a quantitative relationship between defect size, location, and fatigue performance in AM-SFTs. The findings will provide deeper insights into the role of porosity in fatigue behavior and offer more efficient methodologies for fatigue characterization in AM composites.

Keywords Fatigue limit · Engineered defects · AM-SFTs · IRT · Damage

Introduction

Additive manufacturing has revolutionized material design and production, particularly with its ability to fabricate lightweight, high-performance components with tailored properties [2]. Among the most promising materials in this field are additively manufactured short-fiber thermoplastics, which combine polymer matrices with reinforcing short fibers to achieve improved mechanical properties [2]. These materials are commonly produced using fused deposition modeling, a process that offers manufacturing flexibility but can introduce defects, such as porosity, which significantly impact structural integrity under cyclic loading [1, 3].

Porosity in AM components is a critical concern as it acts as a stress concentrator, promoting damage initiation and propagation, thereby accelerating fatigue failure [4]. Fatigue damage, which progressively accumulates in materials under cyclic loading, can substantially affect the residual stiffness and strength of SFTs. Gaining insight into this damage and its influence on material properties is crucial for accurately predicting the lifespan and structural integrity of SFT-based components [1]. Traditional fatigue characterization using stress-life testing is time-consuming and resource-intensive, requiring numerous specimens and long test durations [5]. To address these challenges, infrared thermography has emerged as an effective alternative, offering rapid, real-time monitoring of damage accumulation by detecting thermal responses during cyclic loading [6, 7].

Harsh Gandhi · Pharindra Pathak · Suhasini Gururaja
Department of Aerospace Engineering, Auburn University, Auburn, AL 36849
e-mail: hsg0014@auburn.edu; pzp0057@auburn.edu; suhasini.gururaja@auburn.edu

Garrett Pataky et al. (eds.), *Fracture, Fatigue, Failure, Damage Evolution and Thermomechanics & Infrared Imaging, Vol. 4*, Conference Proceedings of the Society for Experimental Mechanics Series, https://doi.org/10.13052/9788743808305-2

This study investigates the effect of engineered defects on the fatigue behavior of AM-SFTs. The specimens, composed of CF/ABS, were fabricated using FDM with ±45° fiber orientations following the ASTM D638 Type I standard. Engineered defects of 6 mm diameter were introduced in the 8th layer (middle layer) of select specimens to study the influence of porosity on fatigue performance. A total of 15 specimens were tested under static and fatigue loading conditions and monitored using IRT to capture temperature variations corresponding to fatigue damage initiation.

Background

Materials

To investigate the fatigue performance of AM-SFTs with varying defect configurations, a detailed test matrix was developed, as shown in **Table. 1**. The specimens, fabricated according to ASTM D638 Type I standards, were composed of 20% chopped carbon fibers and 80% ABS. The samples were printed using a fused deposition modeling (FDM) process with a 0.4 mm nozzle diameter and a printing temperature of 240°C. A rectilinear infill pattern with 100% sparse infill density was utilized to ensure consistent material properties throughout the specimens. The fine print setting was selected to improve dimensional accuracy and surface quality. The total printing duration for each specimen was approximately 20 min per specimen. The ±45° fiber orientation was chosen to mitigate shear failure observed in previous tests with 0° laminates, which exhibited premature failure due to shear-dominated loading. This orientation helps distribute the load more effectively across the matrix, improving structural integrity during cyclic loading.

Table 1 Test Matrix for 3D Printed Specimens

Specimen	Material	Fiber aspect ratio (L / D)	3D Printer	% Infill	Ratio $\left(\frac{\sigma_{min}}{\sigma_{max}}\right)$	Frequency (Hz)	Nozzle Dia (mm)
ASTM D638 Type I	80 wt.% ABS 20 wt.% Carbon Fiber	5-8	Bambu Lab X1 Carbon	100	0.1	15	0.4
Configuration	**Specimen ID**	**Orientation (deg)**	**Defect Layer**	**No. of Replicates**	**Specimen Printed**	**Testing**	**Testing Completed**
Pristine	PRIS-STC-ALT45-#	±45	N/A	5	5	Static	5
Pristine	PRIS-FTG-ALT45-#	±45	N/A	5	5	Fatigue	5
6 mm Dia on Middle/8th Layer	6D-8L-FTG-ALT45-#	±45	8th	5	5	Fatigue	5
Total Specimens				**15**	**15**		**15**

The ASTM D638 Type I geometry was chosen due to its widespread use in tensile and fatigue testing of polymeric materials, providing reproducible and reliable mechanical property assessments [7]. The specimen geometry, depicted in **Fig. 1**, included a gauge section measuring 57 mm in length and 13 mm in width, with a thickness of 3.2 mm. Each

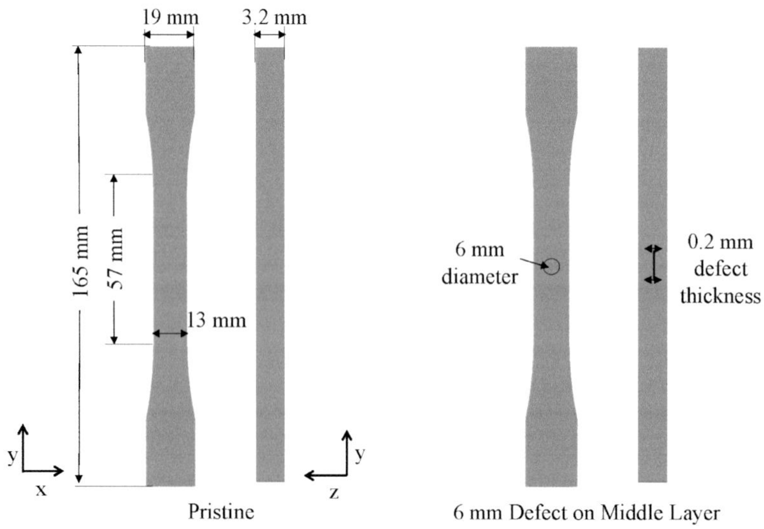

Fig. 1 ASTM D638 Type I Standard Specimen containing 20% CF and 80% ABS materials: pristine specimen (left), defect specimen (right)

specimen consists of 16 layers (0.2 mm thick) additively manufactured on top of each other. As shown in **Table. 1**, two sets of specimens were printed: Pristine specimens and specimens with 'engineered' defects. Engineered defect specimens comprised a 6 mm diameter pore region that was not printed, strategically placed on the 8th (middle) layer.

IRT Testing Procedure

All tests, static tensile and fatigue tests were conducted using an MTS Landmark servohydraulic universal test frame with a 100 kN capacity, as shown in **Fig. 2**. Quasi-static tensile tests were performed at a 2 mm/min displacement-controlled rate to determine the mechanical properties—stiffness and ultimate tensile strength of the CF/ABS composite specimen. Gage pressure was kept around 325 psi to prevent specimen slippage. The strain was measured using an extensometer with a 5 cm gauge length. Five specimens were tested, yielding an average ultimate tensile strength (σ_{uts}) of 27.00 ± 1.49 MPa, failure strain (ϵ_f) of $4.66\% \pm 0.46\%$, and an elastic modulus (E_0) of 2.02 ± 0.16 GPa, which served as the baseline for subsequent fatigue testing.

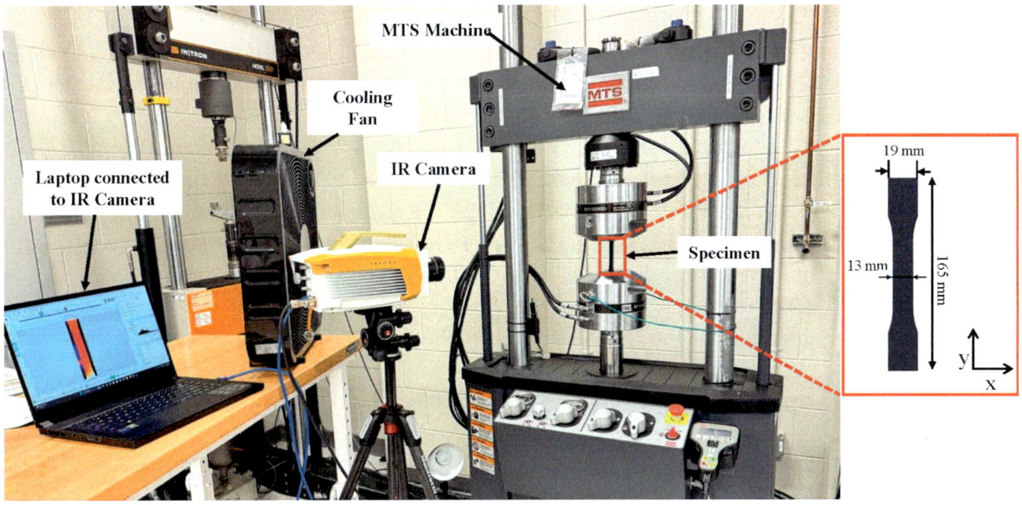

Fig. 2 Experimental Setup for Fatigue Testing

Tension-tension fatigue tests were conducted under sinusoidal load-controlled conditions with a stress ratio $R = 0.1$. The maximum and minimum loads were set as percentages of the average σ_{uts}. The loading frequency was maintained at 15 Hz, following the approach in [8, 9, 10]. **Fig. 3** illustrates the staircase cyclic loading with $R = 0.1$. Temperature

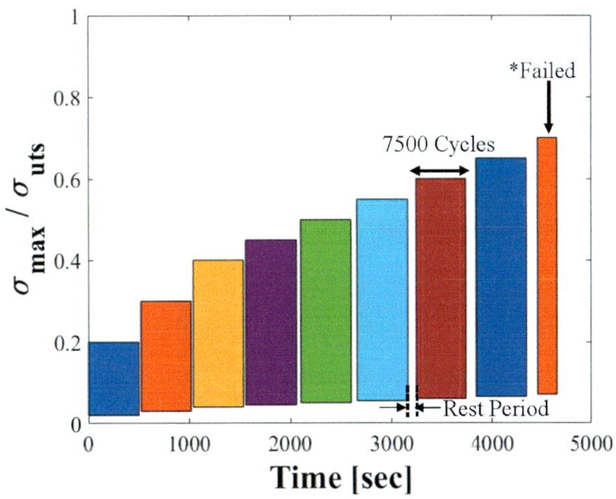

Fig. 3 Staircase loading applied to a specimen to determine fatigue limit rapidly with an early increment of 10% σ_{uts} followed by 5% σ_{uts} increments to σ_{max}. The stress ratio R = 0.1 was maintained throughout the test. The rest period between load steps ensured the specimen cooled down to room temperature after every step load

measurements were obtained in real-time using a Telops Spark M150 MWIR cooled camera (640 × 512 pixels, 20 mK NETD, 3-5.4 μm spectral range) at a data acquisition frequency of 100 Hz. Data acquisition was done via Reveal IR software, and surface temperature evolution was analyzed to identify fatigue limits and correlate them with defect location and fatigue performance. The fatigue limit was determined by plotting the stabilization temperature change (ΔT_{stab}) for each step ($\sigma_{\max}$). The intersection of bi-linear curves fitting the ΔT_{stab} versus $\sigma_{\max}$ data marked the fatigue onset point, identified by the most significant change in slope. The step size (N_{step}) was fixed at 7500 cycles, based on trial tests, to establish the Phase II stabilized steady-state zone. A minimum of five replicates were tested for reliability. Fatigue testing was paused after each load step to allow cooling to room temperature before the next step. The cooling phase called the 'rest period' in **Fig. 3**, ensured specimens started each load step at ambient temperature. Following each rest period, a static test was performed to assess stiffness degradation as a function of load cycles.

Results and Discussion

The fatigue behavior of both pristine and defected specimens was examined through IRT during cyclic loading at increasing stress levels, as shown in **Fig. 4a - 4b**. The pristine specimen displayed minimal temperature variations across all stress levels, with only slight increases at 60% and 65% of the ultimate tensile strength (UTS), indicating a more uniform stress distribution and a delayed onset of internal damage (see **Fig. 4a**). In contrast, the specimen with a 6 mm diameter defect in the middle layer showed notable localized heating at higher stress levels (55%, 60%, and 65% of UTS) (see **Fig. 4b**). The localized heating observed in the engineered defect specimen suggests stress concentrations around the engineered defect, highlighting the role of porosity in accelerating fatigue damage. However, the overall temperature for both pristine and engineered defect specimens remained almost constant, with the defect specimen showing more localized heating due to stress concentration at higher loading levels.

Despite the observed thermal variations, the fatigue limit of the defected specimen, determined through a bilinear plot, was 49.52% ± 0.99% UTS (see **Fig. 5a - 5b**), which was only marginally higher than the pristine specimen's fatigue limit of 49.04% ± 1.28% UTS (see **Fig. 6a - 6b**). This suggests that the defect did not drastically reduce the fatigue resistance as initially expected.

Several factors contributed to this unexpected outcome. Firstly, the MTS machine used in the testing had a load capacity of 100 kN. Since the failure load of the tested specimens was around 1.08 kN, the applied loads were lower than

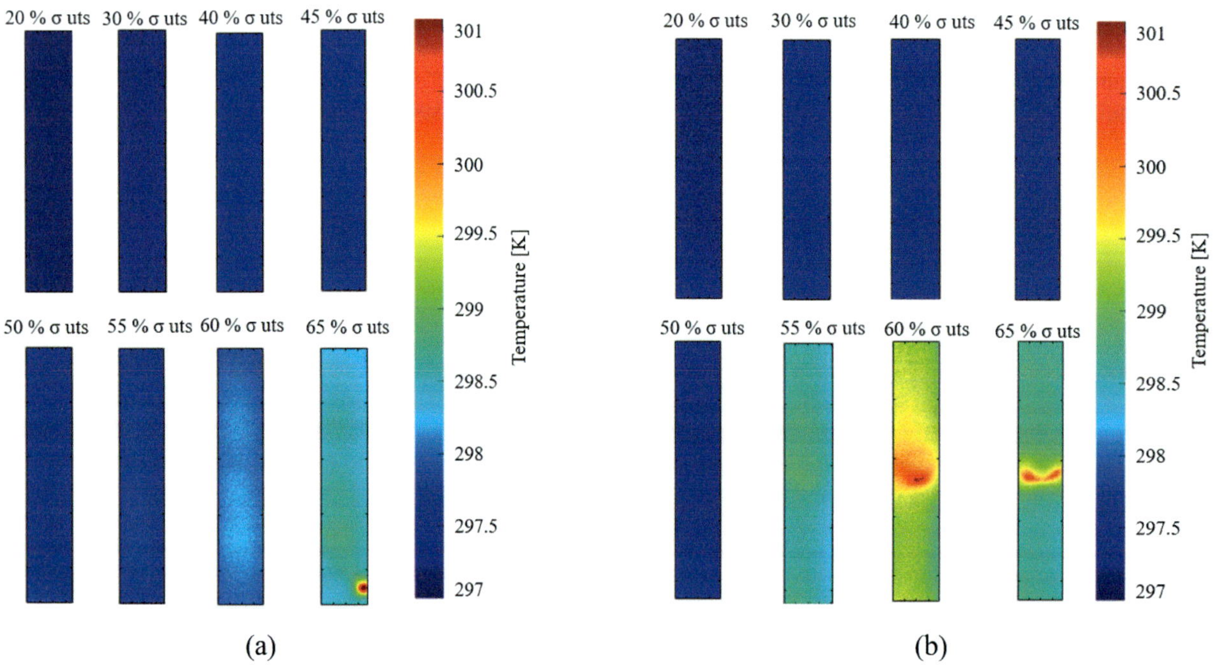

Fig. 4 Temperature displayed within the region of interest for each step loading for both (a) Pristine and (b) Engineered Defect Specimens

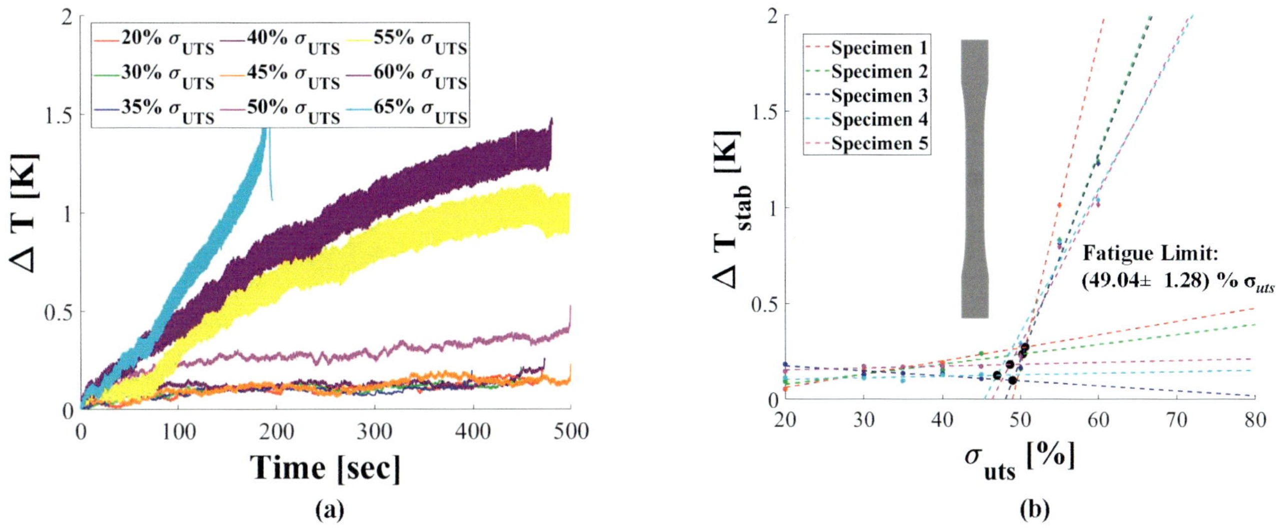

Fig. 5 For Pristine, (a) Temperature vs. Time plot showing the temperature rise for Specimen 4 and (b) Fatigue Limit calculated from the Bilinear Plot for all the specimen

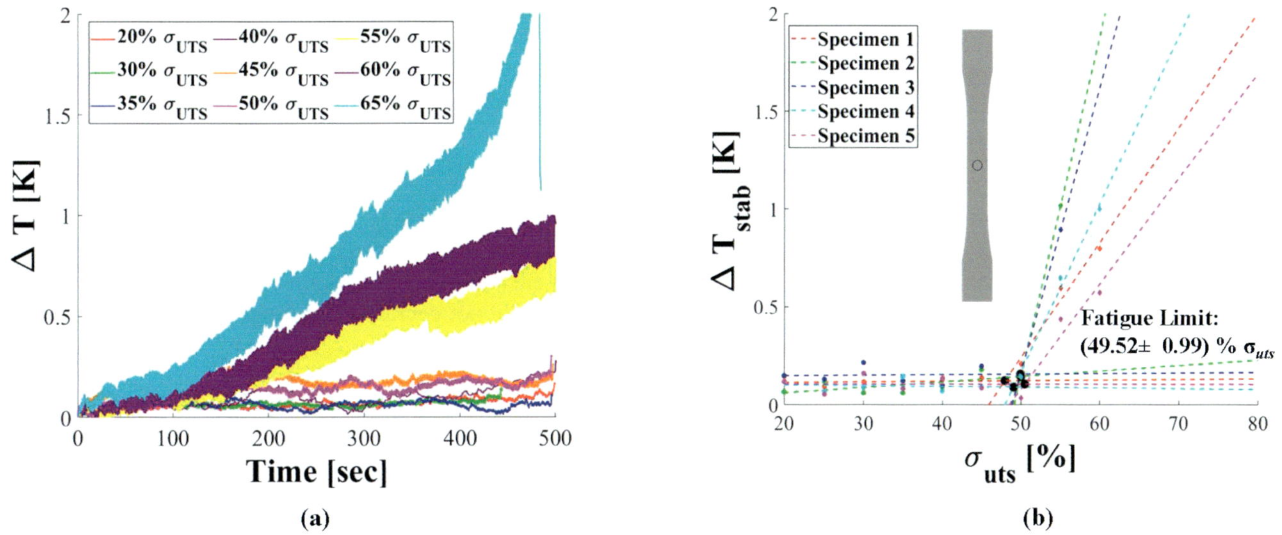

Fig. 6 For 6mm diameter defect on the middle layer, (a) Temperature vs. Time plot showing the temperature rise for Specimen 2 and (b) Fatigue Limit calculated from the Bilinear Plot for all the specimen

the resolution of the MTS machine, potentially affecting the accuracy of the applied load. Secondly, although the engineered defect was intended to be placed in the middle layer (the 8th layer out of 16), micro-CT imaging revealed discrepancies in the actual defect size and shape. The analysis showed that the CF/ABS deposited filament over the defect location, effectively eliminating the void. This indicated that material consolidation during printing contributed to partially filling the defect (see **Fig. 7**), altering its intended geometry, and reducing its influence on fatigue performance. Additionally, due to the defect's depth within the specimen, the thermal signals detected by the infrared camera might have been lower, limiting the ability to capture conclusive temperature data at lower stress levels. This, combined with the resolution constraints of the MTS machine, may have contributed to the inconclusive temperature readings observed at lower load levels. Although localized heating was observed at 55-65% UTS, corresponding to the onset of fatigue damage around the defect, further investigation is needed to determine the defect's true impact on damage initiation at lower stress levels.

In summary, while the presence of the defect was expected to reduce the specimen's fatigue performance due to stress concentration, the results indicate that the fatigue resistance of the defected specimen was comparable to the pristine

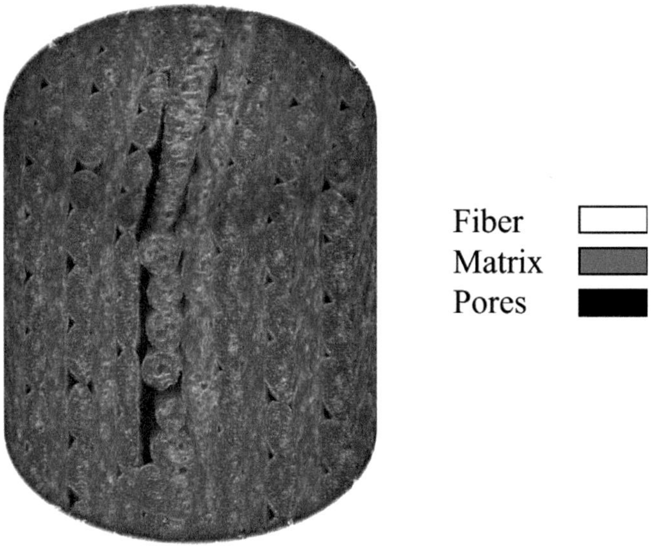

Fig. 7 Microstructure of engineered defect specimen at 0.6 microns voxel size

specimen. This finding emphasizes the influence of testing machine limitations, defect location, and stress redistribution mechanisms in determining the fatigue behavior of specimens with induced defects.

Conclusions and Future Work

This study demonstrated the impact of engineered defects on the fatigue behavior of AM-SFTs through IRT and fatigue limit estimation. Both pristine and engineered defect configurations were considered, with cyclic step-loading applied during the fatigue tests on each specimen. The results showed that while defects were expected to act as stress concentrators and reduce fatigue limits, the observed fatigue threshold for the defected specimen did not decrease as anticipated. This suggests that the chosen defect did not drastically affect the overall fatigue performance under the tested conditions.

Future work will focus on increasing the thickness of the specimen from 16 layers (3.2 mm) to potentially 32 layers (6.4 mm), which will increase the UTS value and ensure that the input loads during step loading remain within the resolution of the testing machine. Additionally, the effects of defect placement will be further explored by assessing the fatigue response of specimens with a 6 mm diameter defect positioned in the third layer. This will provide insights into how defect location influences fatigue life.

By exploring these different configurations, future work will enable a better understanding of the mechanisms behind defect-induced fatigue and help refine the design and manufacturing processes for more reliable AM-SFT components.

References

1. P. Pathak, S. Gururaja, V. Kumar, D. Nuttall, A. Mahmoudi, M. M. Khonsari, and U. Vaidya. Examining infrared thermography based approaches to rapid fatigue characterization of additively manufactured compression molded short fiber thermoplastic composites. Composite Structures, 351:118610, 2025.
2. A. Brasington, C. Sacco, J. Halbritter, R. Wehbe, and R. Harik. Automated fiber placement: A review of history, current technologies, and future paths forward. Composites Part C: Open Access, 6:100182, 2021.
3. C. Colombo, A. Tridello, A. P. Pagnoncelli, C. A. Biffi, J. Fiocchi, A. Tuissi, L. M. Vergani, and D. S. Paolino. Efficient experimental methods for rapid fatigue life estimation of additive manufactured elements. International Journal of Fatigue, 167, Part B:107345, 2023.
4. M. Ivey, G. W. Melenka, J. P. Carey, and C. Ayranci. Characterizing short-fiber-reinforced composites produced using additive manufacturing. Advanced Manufacturing: Polymer & Composites Science, 3(3):81–91, 2017.
5. W. S. de Carvalho, J. Draper, T. Terrazas-Monje, A. Toumpis, A. Galloway, and S. T. Amancio-Filho. Fatigue life as-sessment and fracture mechanisms of additively manufactured metal-fiber reinforced thermoplastic hybrid structures produced via ultrasonic joining. Journal of Materials Research and Technology, 26:5716–5730, 2023.
6. D. Palumbo, R. De Finis, P. G. Demelio, and U. Galietti. A new rapid thermographic method to assess the fatigue limit in gfrp composites. Composites Part B: Engineering, 103:60–67, 2016.

7. Z. Jia, M. L. Pastor, C. Garnier, and X. Gong. Fatigue life determination based on infrared thermographic data for multidirectional (md) cfrp composite laminates. Composite Structures, 319:117202, 2023.
8. Y. Yan, M. L. Pastor, E. Abisset-Chavanne, and X. Gong. Rapid identification of fatigue degradation law of composite material based on ir thermography. Composite Structures, 354:118787, 2025.
9. N. Manoharan, P. Pathak, S. Gururaja, V. Kumar, and U. Vaidya. Rapid fatigue characterization via infrared thermography of am-cm composites. In Springer Nature Switzerland, editor, Challenges in Mechanics of Biological Systems and Materials, Thermomechanics and Infrared Imaging, Time Dependent Materials and Residual Stress, volume 2, pages 53–58. 2023.
10. J. Montesano, Z. Fawaz, and H. Bougherara. Use of infrared thermography to investigate the fatigue behavior of a carbon fiber reinforced polymer composite. Composite Structures, 97:76–83, 2013.

Chapter 3
Fracture Properties Identification using Full-field Measurements: Some Important Concepts and Validation

João Carlos A. D. Filho, Lukas Wittevrongel, Amar Peshave, Pascal Lava, and Fabrice Pierron

Abstract This article studies the effect of Digital Image Correlation uncertainties on fracture parameters obtained from a CT test specimen. It uses synthetic image deformation from finite element simulations to identify such errors. The paper looks at systematic errors from the DIC process as well as the effect of image noise.

Keywords Fracture mechanics · Synthetic image deformation · Uncertainty quantification · Digital image correlation

Introduction

Fracture mechanics is a widely used tool in academia and industry to predict the failure of materials. An important parameter to such models is the fracture toughness in the different modes of fracture, I, II and III. Many techniques have been developed over the years to obtain such parameters, and it is beyond the scope of the present paper to review them. They mostly use the load recorded by the load cell of the test machine together with some (semi-) analytical solutions to derive the fracture properties.

In the last twenty years, full-field optical measurements have gradually spread in the experimental mechanics community, providing very dense sets of kinematic data (typically tens of thousands or more), hence the term 'full-field'. Digital Image Correlation (DIC) is the most widespread [1], with many commercial systems providing turnkey solutions. It is therefore not surprising that many studies have used DIC data to extract fracture parameters, as reviewed in [2]. There are generally two main families of approaches: either fitting the crack-tip displacement field with a Williams series [3] to obtain the stress concentration factor; or using the J-integral [4].

DIC is a complex measurement chain, highly nonlinear and subject to many potential sources of error. As a consequence, while it is easy to process DIC data for stress concentration factors or toughness, it is hard to establish realistic error bars on these quantities. This is the objective of the present contribution. It relies on the technology of synthetic image deformation developed initially in [5, 6] to perform uncertainty quantification in material constitutive model identification with the Virtual Fields Method. It is extended here to the determination of stress concentration factors and J-integrals.

Simulation of DIC measurements on CT specimen

A finite element (FE) model of a Compact Tension (CT) specimen was developed using Abaqus v.6.6, based on the ASTM standard E399 [7]. Plane stress CPS4R linear elements were used. The dimensions and the mesh are reported in Fig. 1. The bottom and top holes were constrained in all translational degrees of freedom, except the top hole vertical displacement which was left free while a vertical load of 35 N was applied. Large transformations were enabled. The material selected for this study is PMMA with a Young's modulus of 3260 MPa and Poisson's ratio of 0.36. The material thickness is 1 mm.

Based on the FE model displacements, a numerically-generated speckle pattern was deformed according to the procedure described in [6]. The pattern was generated with the internal speckle pattern generator in the MatchID software. It consists of

João Carlos A.D. Filho · Lukas Wittevrongel · Amar Peshave · Pascal Lava · Fabrice Pierron
MatchID NV, Leiekaai 25A, 9000 Ghent, Belgium
e-mail: joao.filho@matchid.eu; lukas.wittevrongel@matchid.eu; amar.peshave@matchid.eu; pascal.lava@matchid.eu; fabrice.pierron@matchid.eu

Garrett Pataky et al. (eds.), *Fracture, Fatigue, Failure, Damage Evolution and Thermomechanics & Infrared Imaging, Vol. 4*, Conference Proceedings of the Society for Experimental Mechanics Series, https://doi.org/10.13052/9788743808305-3

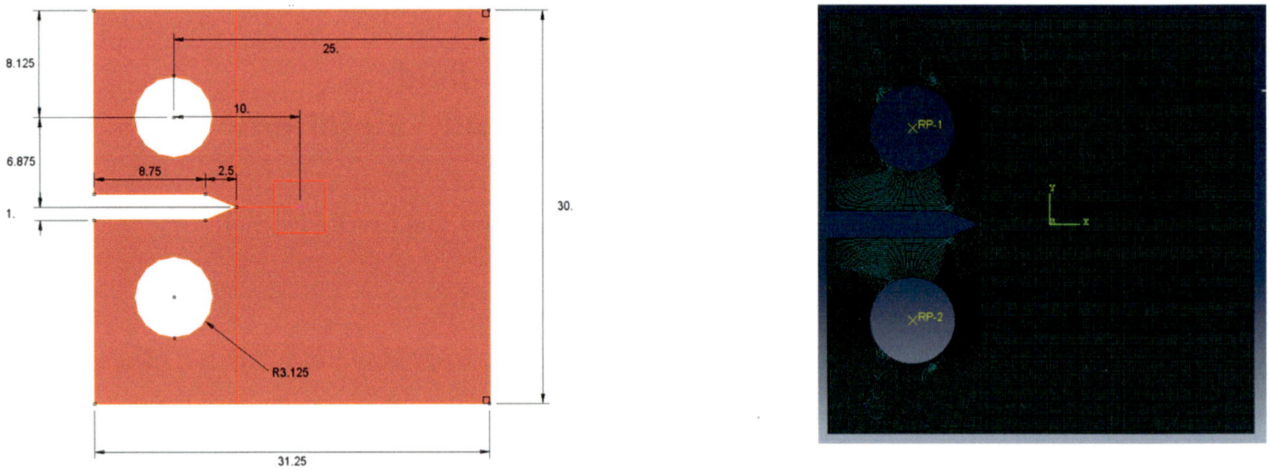

Fig. 1 Dimensions (in mm) and finite element (FE) model of the CT specimen

a regular grid of dots of size about 6 pixels, which centres are slightly moved in a random manner, creating a consistent-size pattern with enough randomness and an isotropic contrast pattern, as seen in Fig. 2. A virtual 8-bit camera of 2448 x 2048 pixels was used to sample the pattern, and deformed images were created for ten load steps from 3.5 to 35 N. The deformed images were then passed through the DIC engine with the parameters recorded in Table 1, according to the recommendations of the DIC guide of good practice [8].

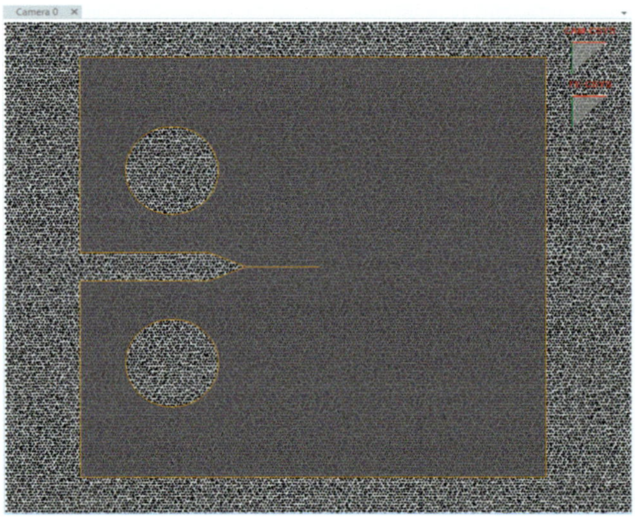

Fig. 2 Image deformation based on the CT FE model

Table 1 DIC parameters

Method	2D DIC
Subset size (pixel)	21
Step size (pixel)	3
Matching criterion	ZNSSD
Shape function	Quadratic
Interpolation	Local bicubic splines
Prefiltering	Gaussian, 5x5 pixels
Strain calculation	Local bilinear polynomial fit
Strain window	3, 5, 7

In the present case, the crack does not grow, the objective was to compare the crack results extracted directly from Abaqus with that from the MatchID crack module and then investigate the impact of the DIC parameters on the results.

The calculation of the stress concentration factor is performed by fitting a 10th order Williams series around the crack tip over the domain visualized in Fig. 3. This is performed in the least-square sense and both K_I and K_{II} are determined. Consistently in this work, because of the test case, K_{II} was found close to zero. For the J-integral, it was calculated over a square path represented in Fig. 3, while the surface J-integral domain is also visible on the same figure, together with the weighing function q which is 0 at the external boundary of the surface and 1 at the internal boundary. It is worth noting that the Williams series fitting is not affected by the choice of the strain window, as it only considers displacements, contrary to the J-integrals. Definition of the contour J-integral and the surface one (also called Equivalent Domain Integral) can be found in [2].

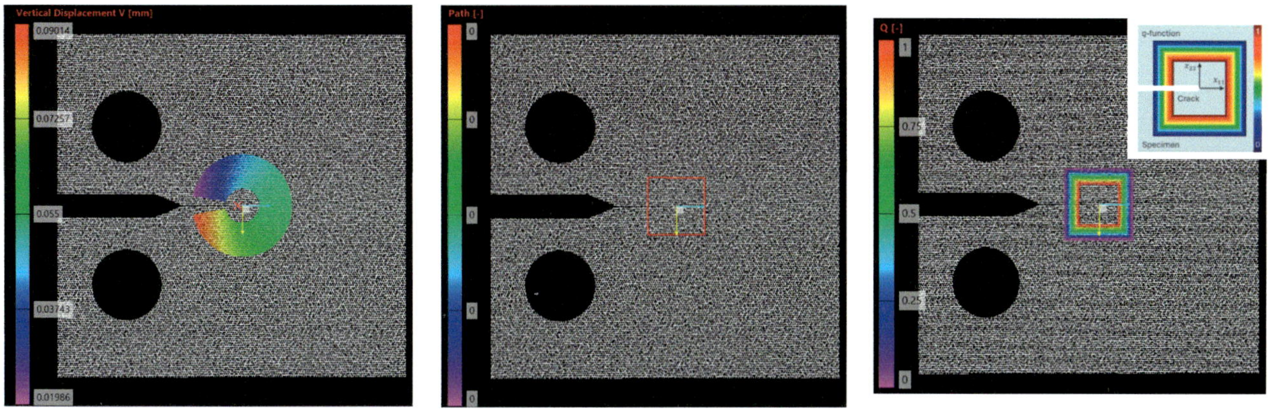

Fig. 3 Williams series domain (left), contour J-integral domain (centre) and surface J-integral domain with the weighing function

Results

The vertical displacements and strains between the pure FE model and the DIC data through image deformation are shown in Figs. 4 and 5 for the last load step at 35 N. The match between the two is clear, though one can see small differences in the strain maps caused by residual interpolation errors in the image deformation process.

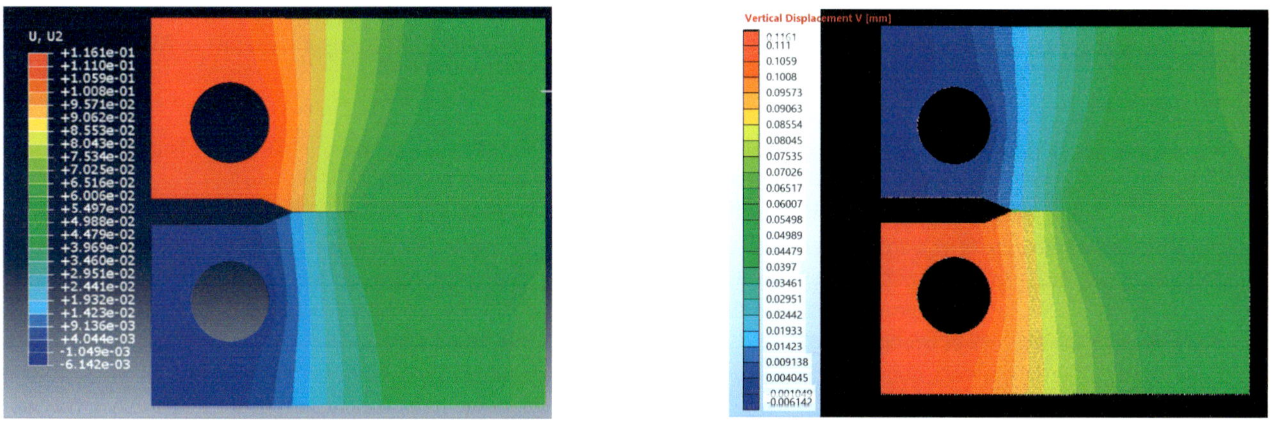

Fig. 4 Vertical displacement maps for the FE model (left) and the synthetically deformed images (right)

The first step consists in a verification of the implementation of the calculations of K and J in MatchID. For this purpose, displacements were directly imported into MatchID and strains calculated as reported in Table 1, with a minimal strain

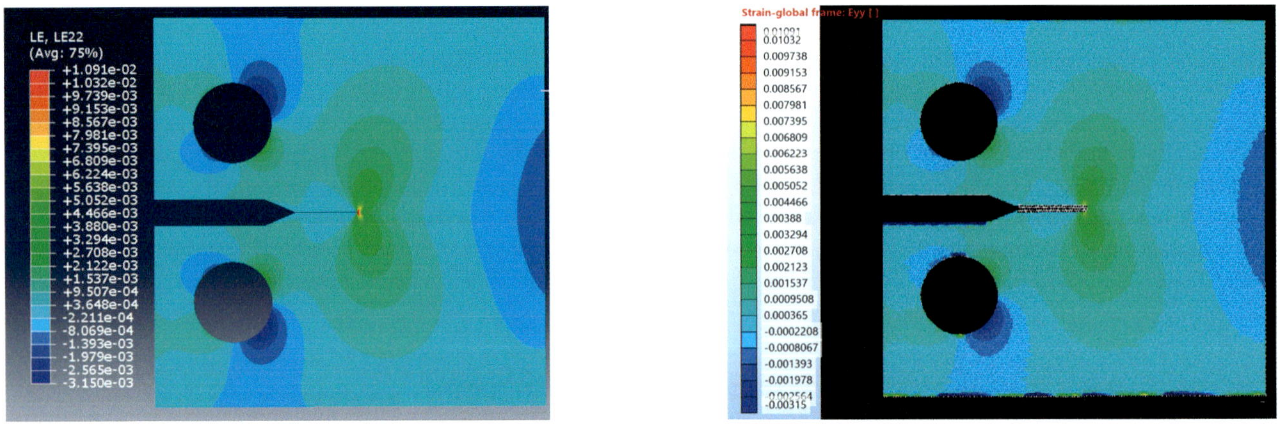

Fig. 5 Vertical strain maps for the FE model (left) and the synthetically deformed images (right)

window of 3. The three fracture metrics (K, J_{path} and J_{frac}) were then compared between the Abaqus and MatchID outputs. The results are provided in Fig. 6.

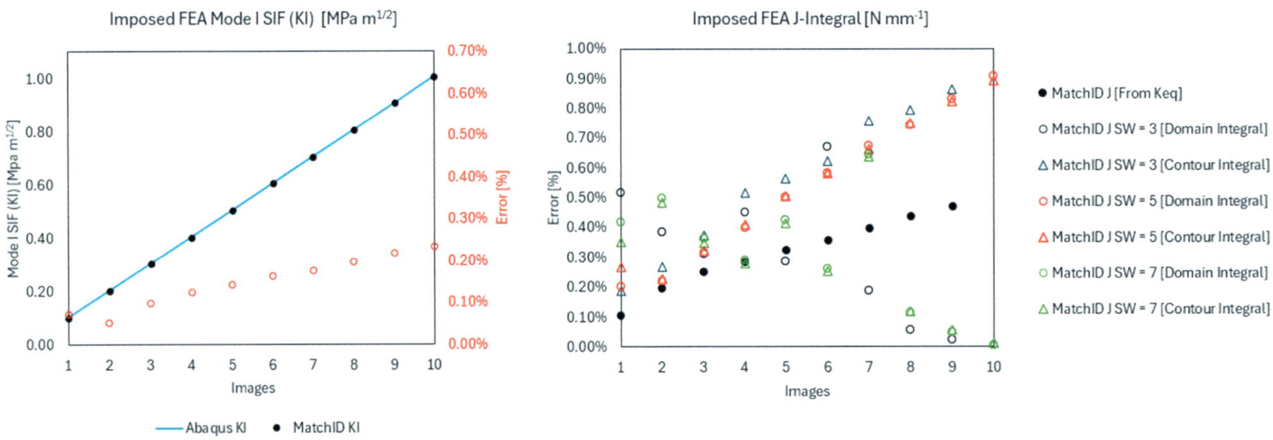

Fig. 6 Verification of the calculation of K from the Williams series (left) and for the J-integrals, for different strain windows. No image deformation, pure FE results loaded into MatchID and strains calculated from displacements in MatchID.

For both K and J-integrals, the differences between Abaqus and MatchID are lower than 1%. There is a limited impact of the strain window too. This result verifies that the MatchID implementation is correct.

The second step is to process the deformed images and compare to the Abaqus results. The data are reported in Fig. 7. For the Williams series, the errors are very similar to that obtained from the pure FE data but there are some larger errors at small loads (small strains) on the path J-integrals. These are caused by edge outliers over the faces of the crack. It is worth noting that the surface integrals do not suffer from these effects as the spurious edge data average out. It is also worth noting that for higher loads, this effect becomes negligible.

Finally, the effect of camera noise was simulated by polluting each image by a Gaussian distribution of noise of zero mean and 1.5 grey level standard deviation for the 8-bit images, a realistic noise level for modern CMOS sensors. The results are summarized in Fig. 8. K is not affected by the camera noise as it bears only on displacements and a large number of points are used for the fit. This is a clear advantage of it. The contour J-integral is the most sensitive to noise as it does not use a lot of (noisy) strain data points, while the surface J-integral is more robust to strain noise. One also sees a reduction of noise with the strain window, as expected. From these results, the advantage of the surface J-integral formulation vs the contour one is clear.

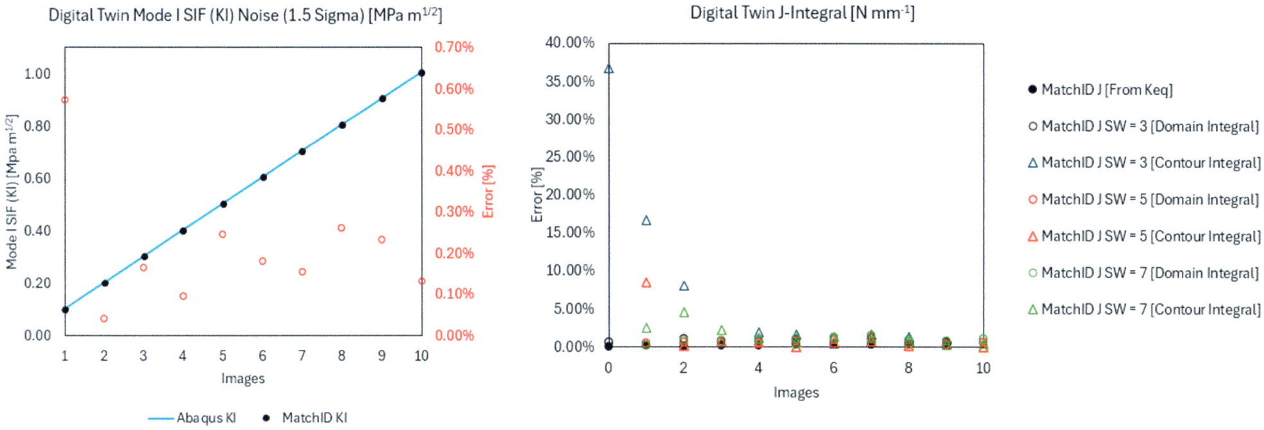

Fig. 7 Error on K (left) and J-integrals (right) for different regularizations (strain windows)

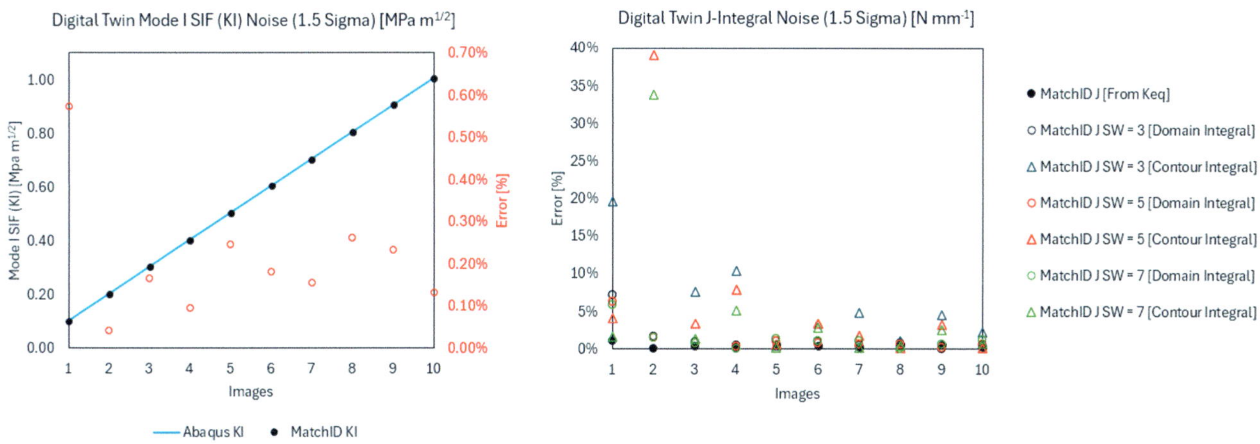

Fig. 8 Error on K (left) and J (right) obtained from noisy images

Conclusion and Future Work

This paper presented a preliminary study on the use of synthetically deformed images to investigate the errors generated by the use of Digital Image Correlation in extracting fracture parameters from fracture tests. The first results are consistent with expectations:

- the formulations in the MatchID software have been verified with the finite element model;
- the Williams series fit to obtain K is much less sensitive to noise in the data as it only uses displacements. However, K only provides an estimate of the fracture toughness;
- the surface J-integral (or Equivalent Domain Integral) is much more robust to strain outliers and noise;

Synthetic image deformation is an ideal tool to investigate systematic and random errors in fracture mechanics parameters obtained from DIC. In the future, effects of imperfect boundary conditions, presence of plasticity at the notch root, crack propagation and other fracture test configurations will be investigated. More realistic DIC errors will also be considered [9].

References

1. Sutton, M.A., J.J. Orteu, and H.W. Schreier, Image correlation for shape, motion and deformation measurements: basic concepts, theory and applications. 2009, New-York: Springer. 1-321.
2. Becker, T.H., Extracting fracture properties from digital image and volume correlation displacement data: A review. Strain, 2024. 60(4): p. e12469

3. Williams, M.L., On the stress distribution at the base of a stationary crack. Journal of theoretical and applied mechanics, 1957. 24(1): p. 109-114.
4. Rice, J., A path-independent integral and the approximate analysis of strain. Journal of applied mechanics, 1968. 35: p. 379-386.
5. Rossi, M. and F. Pierron, On the use of simulated experiments in designing tests for material characterization from full-field measurements. International Journal of Solids and Structures, 2012. 49(3-4): p. 420-435.
6. Rossi, M., et al., Effect of DIC spatial resolution, noise and interpolation error on identification results with the VFM. Strain, 2015. 51(3): p. 206-222.
7. ASTM, E399 - Standard test method for plane-strain fracture toughness if metallic materials. 1990, Amercian Society for Testing and Materials.
8. Jones, E.M.C. and M.A. Iadicola, A Good Practices Guide for Digital Image Correlation. 2018, International Digital Image Correlation Society.
9. Peshave, A., et al., Practical Uncertainty Quantification Guidelines for DIC-Based Numerical Model Validation. Experimental Techniques, 2025. 49: p. 437-457.

Chapter 4
Energy Dissipation Measurement Under Cyclic Torsional Loading Using Infrared Thermography

Daiki Shiozawa, Motoki Yoshiike, and Takahide Sakagami

Abstract In fatigue design, it is necessary to obtain the fatigue strength of the materials used; the conventional method of obtaining the fatigue strength by preparing an S-N curve requires the repeated application of stress to a large number of specimens in the order of 10∧7 cycles, which is enormously time-consuming and costly. A rapid fatigue limit estimation method based on dissipated energy measurements has attracted attention. This method can estimate fatigue limits in a shorter time and at a lower cost than conventional methods. Its applicability to fatigue limit estimation of materials under axial loading has been investigated. In this study, fatigue limit estimation based on dissipated energy measurements was carried out for SCM435 subjected to cyclic torsional loading to investigate whether the method can be applied to torsional loading as well. A compact torsional fatigue testing machine and a microscopic measurement system were constructed. Visible-infrared synchronous measurement was applied to correct the position of torsional deformation. Fatigue limit estimation based on the dissipated energy under torsional loading obtained by the developed fatigue testing machine and measurement system was found to be in good agreement with the fatigue limit obtained from the *S-N* curve.

Keywords Infrared thermography · Energy dissipation · Torsional fatigue

Introduction

The dissipated energy is measured by infrared thermography, and the fatigue limit of the material can be estimated from the stress amplitude at which the dissipated energy increases rapidly [1–4]. Compared to conventional methods, this method can estimate the fatigue limit using a smaller number of test pieces and on the order of 10^4 cyclic loadings, which is expected to reduce time and costs. Although this method has been widely applied to axial loads, there have been few cases of its application to torsional loads. In order to prevent fatigue failure of engines, turbines, motors, crankshafts, and other items that are subjected to torsional loads, it is necessary to design them taking into account the fatigue limit against torsion. Therefore, if this method could also be applied to torsional loads, it would be extremely beneficial for the industrial field. In this study, the dissipated energy of SCM435 under torsional fatigue loads was measured by infrared thermography camera and the fatigue limit was estimated based on the dissipated energy.

Experimental Setup and Specimen

The geometry of the specimen used in this experiment is shown in Figure 1. The material is chromium molybdenum steel SCM435. Since infrared thermography camera cannot capture images of curved surfaces, a flat section with a width of 1 mm and a length of 4 mm was provided in the center of the specimen. A stir case like stress level test was performed on this specimen. In the stir-case like stress level test, a sinusoidal waveform with a stress ratio of R = -1 and a frequency of f = 5 Hz was applied to the specimen, and the shear stress amplitude was increased in a stepwise manner by 20 MPa each time the number of load cycles reached 1000 cycles. At each stress amplitude, the flat part of the specimen was measured using an infrared thermograph camera and a visible camera, respectively, at the time when the number of load cycles reached 300 and

Daiki Shiozawa · Motoki Yoshiike · Takahide Sakagami
Department of Mechanical EngineeringKobe University, Kobe, Japan
e-mail: shiozawa@mech.kobe-u.ac.jp; 248t368t@stu.kobe-u.ac.jp; sakagami@mech.kobe-u.ac.jp

Garrett Pataky et al. (eds.), *Fracture, Fatigue, Failure, Damage Evolution and Thermomechanics & Infrared Imaging, Vol. 4*,
Conference Proceedings of the Society for Experimental Mechanics Series, https://doi.org/10.13052/9788743808305-4

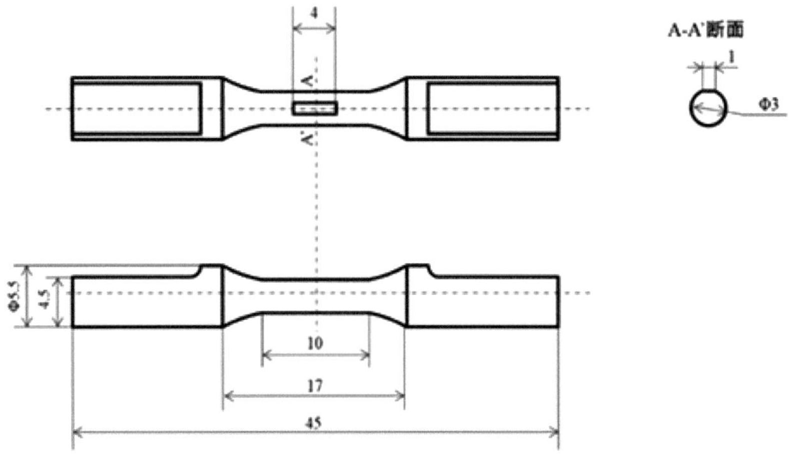

Fig. 1 Geometry of specimen

600 cycles. A microscope lens was attached to the infrared thermography camera to perform magnified measurements of the flat part. The spatial resolution at this time was approximately 20 μm/pixel. The flame rate was 300Hz, and the shooting time was 10s, and the dissipated energy, which is 2nd frequency components of the load frequency, was measured. A position correction process was performed to infrared thermography images using visible images to remove apparent temperature fluctuations that occur when the observation point moves due to deformation or displacement of the specimen [5]. In the position correction process, digital image correlation is performed on the visible images to measure the displacement of the specimen.

Spot markers were painted on the test specimen. The displacement information obtained here was reflected in the infrared image and position correction process was performed. In the constant stress amplitude test, a fatigue test was performed under loading conditions of R=-1 and f=5Hz using a specimen without a flat surface, and an *S-N* curve was obtained. The runout number of cycles at which no fracture occurred was set to 10^7 cycles.

Results

Figure 2 shows the *S-N* curve obtained from the constant stress amplitude test. It was found from Fig. 2 that the fatigue limit of this specimen τ_w was between 300 MPa and 320 MPa, since no fracture occurred at τ_a= 300 MPa.

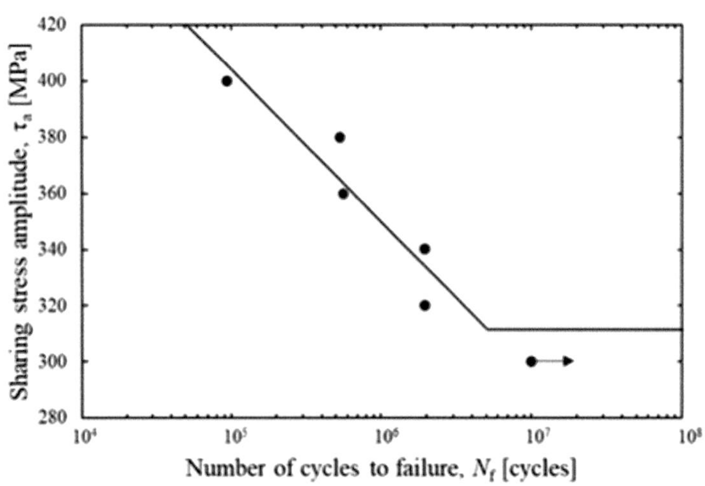

Fig. 2 *S-N* curves

The change in dissipated energy in the stepwise stress increase test is shown in Fig. 3. In Fig. 3, the evaluation area is the flat area in the center of the parallel part of the specimen, and the average value of the data measured twice at each stress amplitude is used as the representative value. It was found from Fig. 3 that the dissipated energy remains constant at stress amplitudes smaller than 320 MPa, and it increases from τ_a =340MPa. For these measurement results, the data for each stress amplitude was divided into two groups, a low-temperature group and a high-temperature group, so that the residual in the linear approximation was minimized, and the boundary value of the grouping was used as the estimated fatigue limit. The estimated fatigue limit τ'_w in this experimental results was 330 MPa. The estimated fatigue limit τ'_w was close to the fatigue limit obtained from the *S-N* curve.

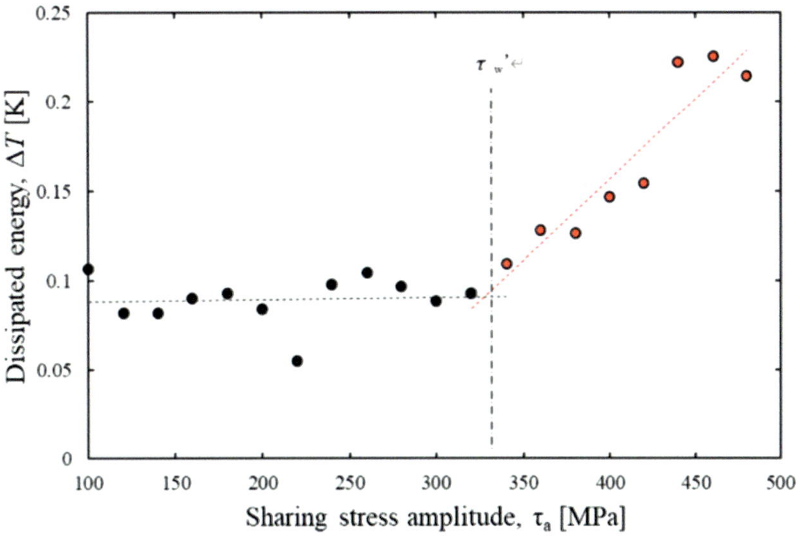

Fig. 3 change in dissipated energy during the stir-case like stress level test.

Conclusion

In this study, the fatigue limit of SCM435 was estimated based on dissipated energy measurements in torsional fatigue. It was found that the fatigue limit was between 300 and 320 MPa from the *S-N* curve obtained from a constant stress amplitude test. From a stir-case like stress level test, it was found that the dissipated energy showed a constant value below a certain stress amplitude, and then it rapidly increased. This estimated fatigue limit at which the dissipated energy rapidly increased was close to the actual fatigue limit.

References

1. Bremond, P., and Potet, P., (2001) Lock-in thermography: a tool to analyze and locate thermomechanical mechanism in materials and structure, *Proc. Of SPIE*, 4360: 560–566.
2. Luong, M. P. (1995) infrared thermographic scanning of fatigue in metals, *Nuclear Engineering and Design*, 158: 363-376.
3. Rosa, G. La and Risitano, A. (2000) Thermographic methodology for rapid determination of the fatigue limit of materials and mechanical components, *Int. Jour. Fatigue*, 22, 1: 65–73.
4. Krapez, J. –C. and Pacou, D. (2002) Thermography detection of damage initiation during fatigue tests, *Proc. of SPIE, Thermosense XXIV*, 4710: 435–449.
5. Uchida Y. Shiozawa D., Hori M., Kobayashi K., Sakagami T., (2022) Advanced Technique for Thermoelastic Stress Analysis and Dissipation Energy Evaluation Via Visible-Infrared Synchronous Measurement, Experimental mechanics, 62:459-470.

Chapter 5
On the Calculation of the Initial Temperature Slope: Key Considerations for the Fatigue Limit Estimation

Mohammad Zaeimi, Ali Mahmoudi, Rosa De Finis, Davide Palumbo, Michael M. Khonsari, and Umberto Galietti

Abstract Over the past four decades, infrared thermography has become a popular non-destructive technique for identifying thermal signature associated with fatigue of material. In this work, by conducting experimental stepwise tests on SS316 samples and employing a unique dataset, a comparison is performed in terms of fatigue limit estimation between different thermal indices, including the mean temperature rise, slope of the mean temperature, first and second harmonic amplitudes of temperature and slope of the mean temperature in cooling stage. However, the main contribution of the work is proposing key considerations in calculating the initial slope of temperature to the purpose of better accuracy and reliability in estimating the thermal fatigue limit from temperature data. It is shown that the time interval and thermal acquisition can affect its measurements.

Keywords Thermographic method · Rapid fatigue estimation · Fatigue assessments · Thermal index · Initial slope of the temperature

Introduction

Investigating the fatigue phenomena offers crucial information for choosing effective materials because the majority of structures experience cyclic loadings throughout their life [1]. The cyclic nature of the fatigue loading creates dislocations, which in turn leads to plastic deformation. Depending on the loading level, it causes microscopic flaws [2–4] to grow and build into macroscopic cracks, and by propagation of macroscopic crack, fatigue failure happens [5].

Over the years, infrared thermography has shown to be a useful and adaptable instrument in the field of experimental mechanics. It enables thorough investigations of structural integrity [6], the defect detection in different materials (steel plates [7], composite structures [8], additively manufactured [9]) and welded joints [10], the study of fatigue cracks [11]), the stress analysis [12] and fatigue assessment [13–15].

In fatigue assessments using infrared thermography, rapid tests were initially designed in the work of La Rosa and Risitano [16] to cut down on testing time and costs. This methodology involves the application of a sequence of incremental loading blocks with constant amplitude leading the material to be subjected to various damage regimes within the loading levels [16].

Following this work, several researchers proposed distinct rapid methods for fatigue limit prediction under cyclic loading which can be categorized in three different groups [13]: (i) direct temperature assessment, (ii) thermoelastic stress analysis and (iii) energy-based approaches. The first one focuses on measuring the mean temperature increase [16] and analyzing

Mohammad Zaeimi · Davide Palumbo · Umberto Galietti
Department of Mechanics, Mathematics and Management, Polytechnic University of Bari, Via Orabona, Bari 70126, Italy
email: mohammad.zaeimi@poliba.it; davide.palumbo@poliba.it; umberto.galietti@poliba.it

Ali Mahmoudi · Michael M. Khonsari
Department of Mechanical and Industrial Engineering, Louisiana State University, 3283 Patrick Taylor Halkl, Baton Rouge, LA 70803, USA
email: alimahmoudi@lsu.edu; khonsari@lsu.edu

Rosa De Finis
University of Salento, Department of Engineering for Innovation, Piazza Tancredi, 73100, Lecce, Italy
email: rosa.definis@unisalento.it

Garrett Pataky et al. (eds.), *Fracture, Fatigue, Failure, Damage Evolution and Thermomechanics & Infrared Imaging, Vol. 4*, Conference Proceedings of the Society for Experimental Mechanics Series, https://doi.org/10.13052/9788743808305-5

the initial slope of the temperature [17]. The second one includes the loss of adiabatic conditions through thermoelastic phase analysis [18] and the deviation from linearity in the first amplitude harmonic [19, 20]. Furthermore, the last one explores fatigue behavior through second amplitude harmonics [21], specific heat loss [22], and entropy-based methods [23]. However, the fatigue assessment using the thermographic technique was proposed to be assessed during a static test, by studying the end of the thermoelastic effect [24].

Interested readers are referred to the work by Zaeimi et al. [13] for a more comprehensive review of all thermography-based approaches for fatigue limit estimation, and the work by Colombo and Vergani [25] and Ricotta et al. [26] for comparing different thermal indices in fatigue limit estimation.

In this work, as the main contribution, the challenges associated with measuring one of the thermal indices, the initial slope of temperature, are discussed. Moreover, five thermal indices including the mean temperature rise, slope of the mean temperature, first and second harmonic amplitudes of temperature and slope of the mean temperature in cooling stage are compared using the same dataset.

Next sections are organized as follows: First, the details about the procedure and experimental setup for rapid evaluation of fatigue limit are provided. Next, the data are prepared and the results are proposed. Finally, the last section concludes the study.

Experiments

Three Stainless Steel 316 (SS 316) samples were used in stepwise tests to prepare essential data for comparing the effectiveness of various thermal indices in fatigue limit prediction. Table 1 and Figure 1 show the material characteristics and geometry of a cylindrical dog bone sample, respectively. In Table 1, ρ is the density, C_p, α and k are respectively the specific heat at constant pressure, thermal expansion and conductivity coefficients; ϵ is the emissivity factor. E, σ_y and σ_u are Young's modulus, yield strength and ultimate strength, correspondingly.

Table 1 Material properties of SS 316 [27]

ρ (kg/m^3)	C_p $(J/kg$ ℃$)$	k $(W/m$ ℃$)$	α (1/℃)	ϵ	E (GPa)	σ_y (MPa)	σ_u (MPa)
7900	490	16.3	17.2e-6	0.98	171	335	640

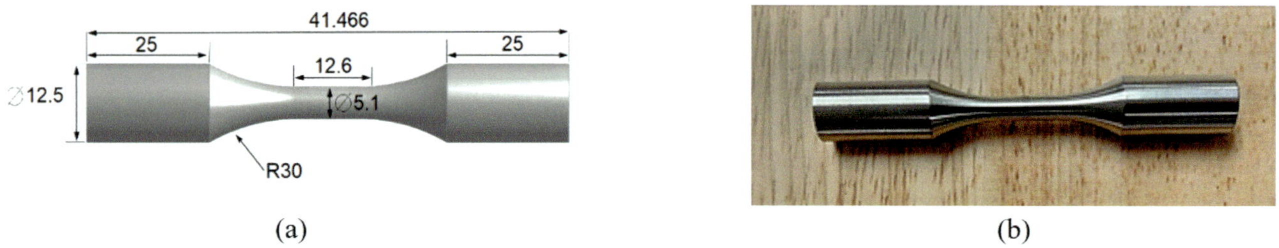

Fig. 1 (a) cylindrical dog bone geometry (in mm) and (b) fabricated sample

As shown in Table 2, a sequence of 12 incremental loading stages was used in the stepwise experiments. Every step is completed with the stress ratio $R = -1$ and a fixed loading frequency of 11 Hz. According to Table 2, the stress amplitude (σ_a) is raised until the material fails after around 20,000 cycles for each stage.

As shown in Figure 2, two thermal cameras and a servo hydraulic fatigue machine (MTS model 370) with a 100 kN loading capability were used for the experiments. In order to observe temperature stability, thermographic data was collected in a single sequence until the end of the test using a micro bolometric infrared camera FLIR A20 (160x140 pixels with a thermal sensitivity NETD $<$ 50 mK). Moreover, using the second camera, a cooled infrared camera X6540 with 640x512 pixels and a thermal sensitivity of NETD $<$ 20 mK, three thermographic sequences were obtained at a higher frame rate.

To remove the impact of ambient temperature on the mean surface temperature, a dummy-unloaded specimen composed of the same material was utilized as a temperature reference. To improve the surface roughness, the gage section was polished using sandpaper and moved through 480, 800, 1200, and 1500 grit sizes. Additionally, to increase the precision of the thermal signal, a matte black coating was added.

Table 2 Loading blocks for stepwise test

Loading block	1	2	3	4	5	6	7	8	9	10	11	12
% σ_u (MPa)	39.8	41.4	43	44.5	46.1	46.9	47.7	48.4	49.2	50	52.3	54.7
σ_a (MPa)	255	265	275	285	295	300	305	310	315	320	335	350

Fig. 2 Stepwise test setup

Thermographic acquisitions are carried out during the test utilizing an IR camera X6540 at 213 Hz for 10 seconds (2130 frames obtained) and a micro bolometric IR camera, FLIR A20, at 5 Hz for the same duration (50 frames acquired). Following a predetermined cycle count, the thermal sequence during each loading phase is recorded. Four sub-steps of thermal measurements were carried out for every loading phase, including 20,000 cycles. The data collected between 1000 and 1110 cycles, 5000 and 5110, 10000 -10110 cycles, and 15000-15110 cycles, respectively. The gauge length are between the extensometer's clips is considered as the region of interest (ROI).

The mean and harmonic components of a thermal signal can be obtained by processing acquired thermal data using a technique known as lock-in thermography [28]. The Fast Fourier Transform (FFT), which allows the temperature signal to be converted into its frequency-domain representation, requires a reference signal (load or strain) in order to process the temperature field. By modeling the temperature as follows, it is possible to obtain all of the temperature components [19]:

$$T(t) = T_0 + \Delta T f t + T_1 \sin(\omega t + \varphi_1) + T_2 \sin(\omega t + \varphi_2) \tag{1}$$

Where the temperature signal $T(t)$, f is the frequency of loading, T_0 is the temperature at the beginning, ΔT is the mean rise, T_1, φ_1, T_2, φ_2 are the amplitudes and phases of the 1st and 2nd Fourier components, respectively. Figure 3 (b) shows a schematic representation of thermal signal processing. As shown in Figure 3 (b), the chosen thermal indices in this work

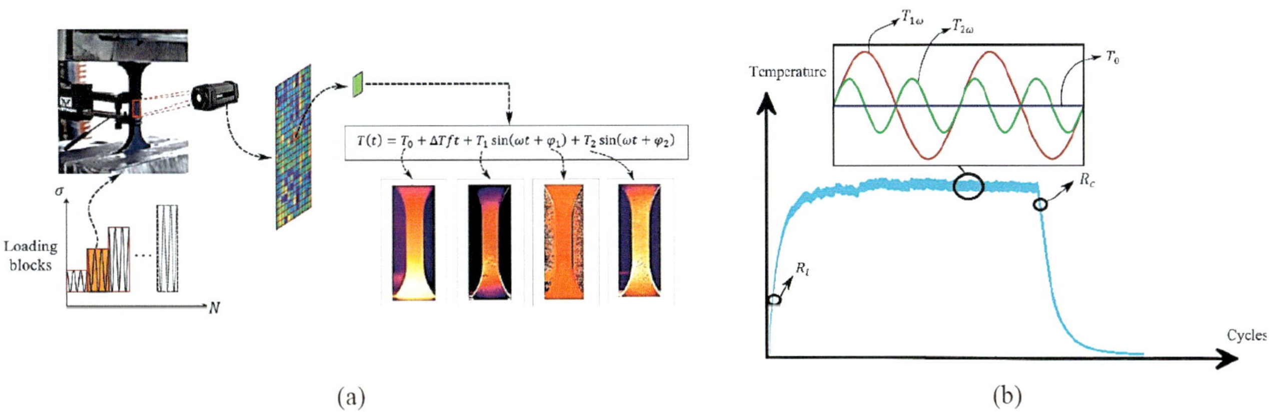

Fig. 3 (a) temperature components acquisition (b) a schematic representation of thermal indices

are: stabilized mean temperature (ΔT), initial slope of the mean temperature (R_i), first amplitude harmonic of temperature ($T_{1\omega}$), second amplitude harmonic of temperature ($T_{2\omega}$) and slope of the mean temperature in cooling stage (R_c).

A number of sequential procedures were engaged in the thermal data processing. First, $T(t)$ was obtained using an infrared thermal camera with a mm/pixel ratio of 0.116 mm. After that, IRTA® software was used to define a particular ROI, and temperature data over time was retrieved for further analysis. For every frame in the time sequence, the average temperature value ($T_0(t)$) for this ROI was calculated. Next, Discrete Fourier Transform (DFT) was used to obtain harmonic components [29].

Results

The overall thermal behavior of samples at various loading levels is explained below. The average surface temperature change during the stepwise test obtained by the micro bolometric camera is shown in Figure 4. After each loading block, a cooling period of about a minute start after temperature stabilization and a specific number of cycles. After that, this process is repeated again until sample fails. Very high temperature, between 160 and 200 °C, was recorded at failure which can be due to the ductility of the material and work hardening.

The thermal footprint is shown for the 3rd sample in Figure 5. Steps 1, 3, and 5 show a slight increase in temperature below the fatigue limit of 306 MPa [27], while steps 9 and 11 show a large increase, suggesting that damage accumulates until the specimen fails (step 11 is the step before the failure of the sample).

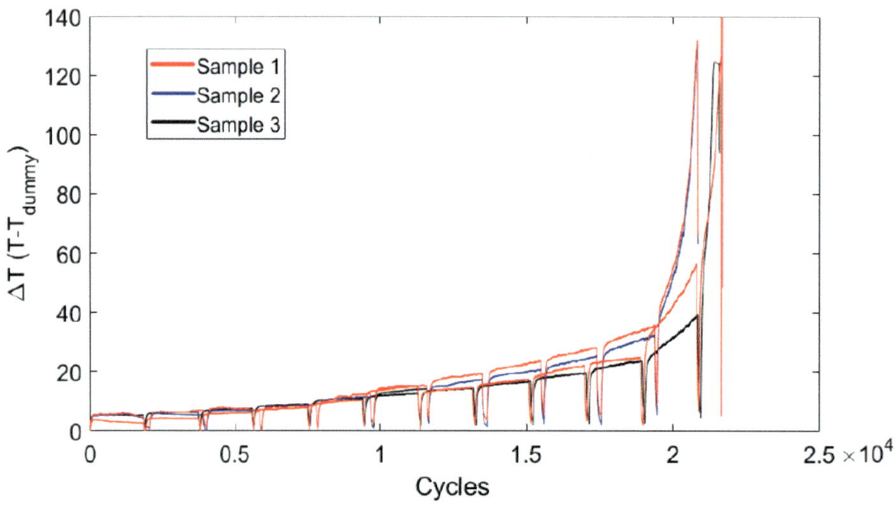

Fig. 4 Temperature trend obtained by a cooled camera for all samples

Figure 6 shows the trend of ΔT and R_c with stress level which was extracted using thermal data from the uncooled camera. ΔT was calculated by subtracting the mean surface temperature of the sample from that of the dummy sample. In the meantime, the gradient of the mean surface temperature during the cooling stage was calculated to determine R_c. Furthermore, the data obtained from the cooled camera was used for $T_{1\omega}$ and $T_{2\omega}$ since a higher sampling rate is required to reconstruct the temperature signal. Figure 7 provides these measured values.

Unlike the other indices, only for R_i, an analysis was conducted to check how the measurement of R_i can be influenced by sampling rate of thermal camera (cooled or uncooled camera) and the time period over which the fitting is applied to thermal data. Within time durations ranging from 1 to 10 seconds, linear fitting was used for the thermal points. It should be noted that the slope was determined using the data points where the temperature begins to climb and shows cyclic trend.

Figure 8 and Figure 9 propose the results only for sample 3 as an example. It is evident that curves from an uncooled camera show more obvious fluctuations, specifically for time duration of 1 to 3 seconds, that prevent the fatigue limit from being estimated from the datapoints, in contrast to curves found through fitting on the data from a cooled camera.

The trends from the cooled camera shows instability at 1 second but stabilizes for longer time duration, whereas for the uncooled camera, notable variations are clear for 1 to 2 seconds.

This analysis highlights the significance of taking precise measurements of the initial slope of the mean temperature. Therefore, it is recommended to use a cooled camera (or a large number of data points) instead of the uncooled one to have

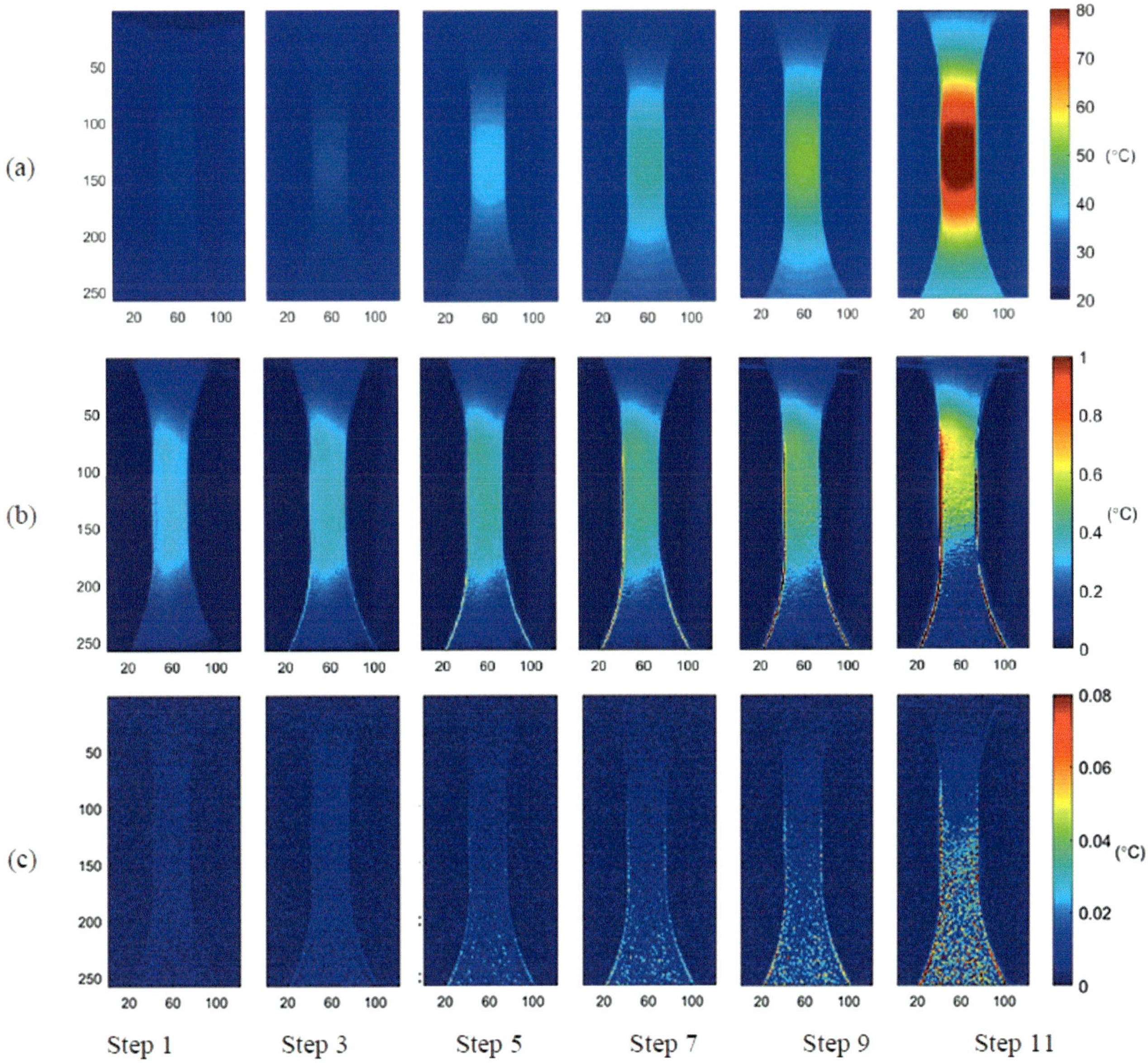

Fig. 5 Variation of thermal signature during stepwise test (sample 3): (a) ΔT (b) $T_{1\omega}$ (c) $T_{2\omega}$

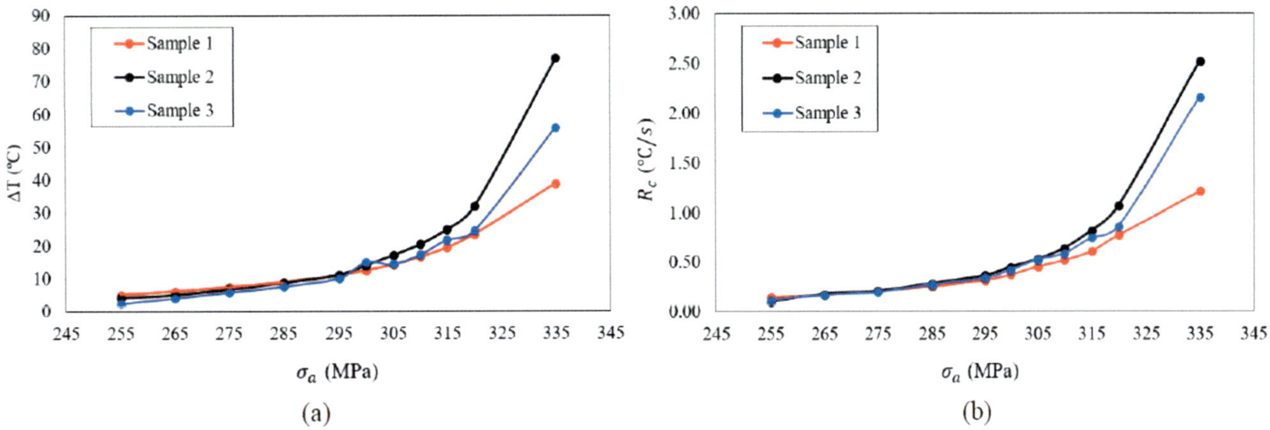

Fig. 6 (a) stabilized temperature rise, and (b) the slope of temperature during cooling stage

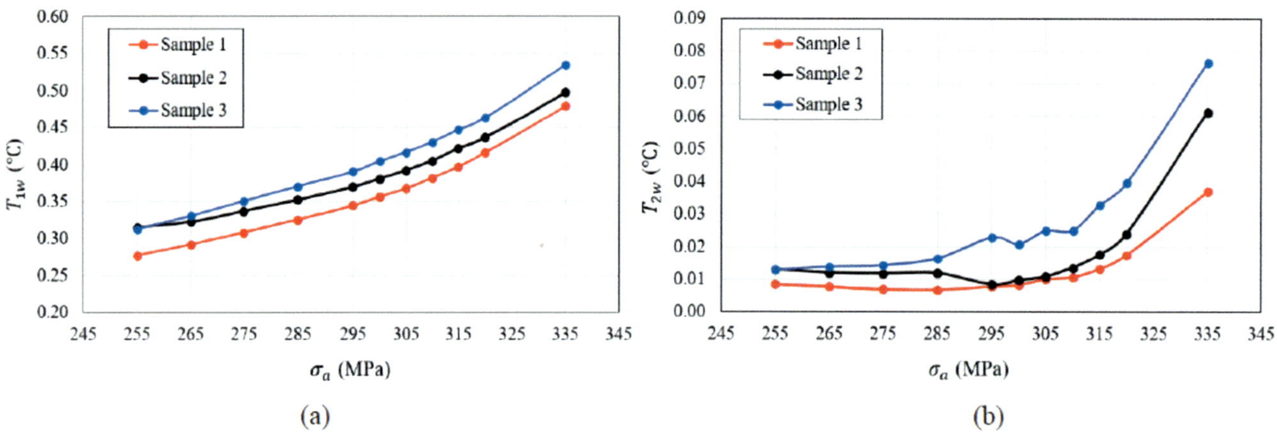

Fig. 7 (a) 1st amplitude harmonic and (b) 2nd amplitude harmonic of the temperature

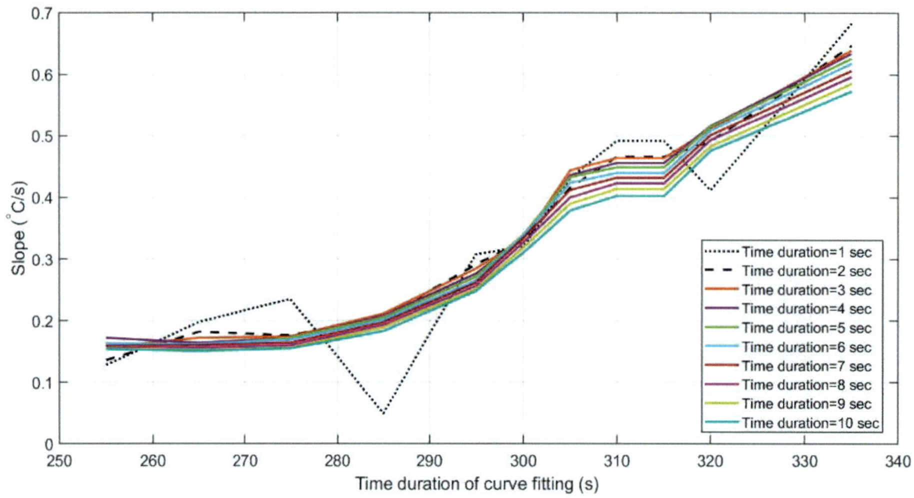

Fig. 8 Rising slope of temperature from linear fitting on uncooled camera data (sample 3)

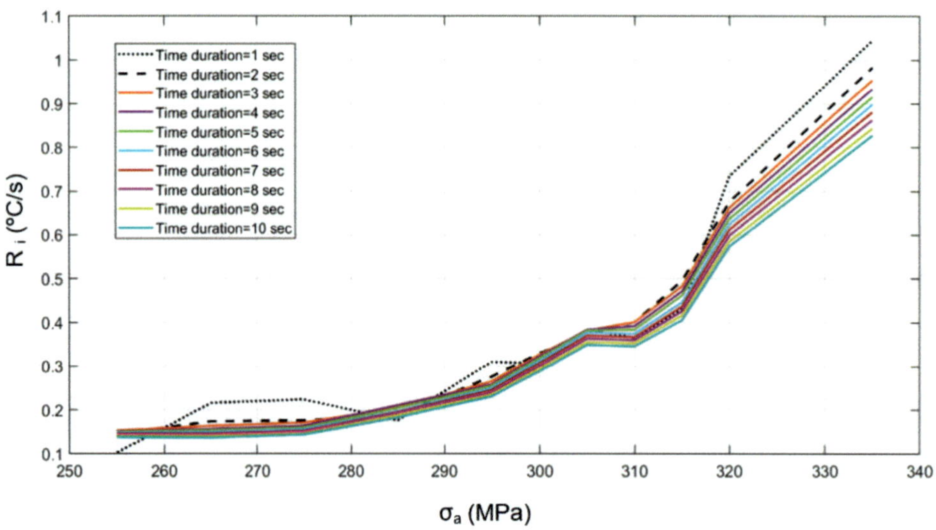

Fig. 9 Rising slope of temperature from linear fitting on cooled camera data (sample 3)

smoother trends. On the other hand, if an uncooled camera is the only choice, it is recommended that curve fitting be done for time durations longer than 2 seconds.

Furthermore, the accuracy of cooled and uncooled camera in fatigue limit prediction is presented below.

Not only for R_i but also for other indices studied here, Loung's method [30] was applied to estimate the fatigue limit.

It involves fitting two regression lines to data points in each thermal index trend that are clearly below and above the knee point [30]. The fatigue limit can be reliably approximated by finding the point where these two lines connect.

Unlike other indices which depends only on the stress level, R_i is also dependent on sampling rate as well as the time duration used for fitting. Thus, the fatigue limit predicted by R_i needs through investigation, as shown in Figure 10. The reference fatigue limit (derived from the S-N curve, 306 MPa [27]) is shown by the dashed red line. It is evident from the figure that, the predicted values from the cooled camera (yellow bars) are consistently closer to the reference value for durations of three seconds or more. Conversely, the estimated values for uncooled camera typically display lower predicted fatigue limits, which are farther away from the dashed line.

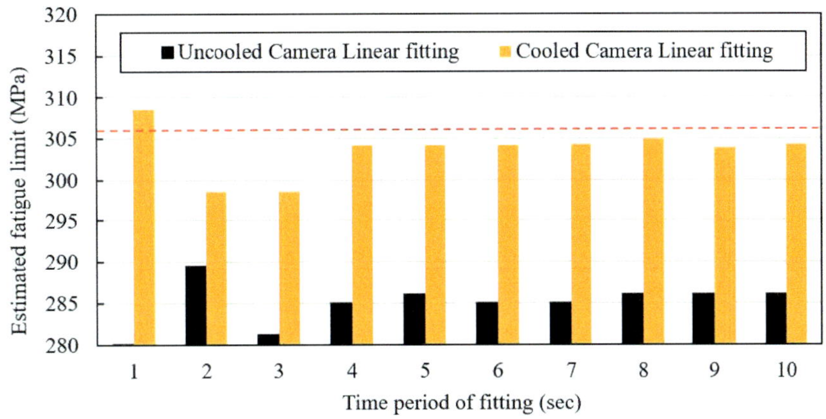

Fig. 10 Fatigue limit predicted by $\boldsymbol{R_i}$ (sample 3)

In light of this finding, the average of the predicted fatigue limits derived from all time periods was used to compare R_i with other thermal indices.

Table 3 shows the fatigue limit estimated using all considered thermal indices in this work. For each index, the discrepancy between the mean of the estimated values from 3 samples and the reference one is also included in the last column of Table 3.

According to the first column, $T_{2\omega}$ and R_i measured from the data acquired by a cooled camera show better performance and accuracy than the others, both showing only a 0.1% difference.

Estimation by ΔT ranks second with a difference of 1.6%. Other methods, such as $T_{1\omega}$, R_c, and R_i from uncooled camera, exhibit errors ranging from 2.1%, 2.9% and 4.4%, respectively.

Table 3 Predicted fatigue limit using Loung's method

Rank	Method	Predicted fatigue limit (MPa)				Difference* (%)
		Sample 1	Sample 2	Sample 3	Average	
1	$T_{2\omega}$(℃)	305.69	304.33	307.21	305.75	0.1
1	R_i (℃/s) from linear fitting and cooled camera	-	308.05	303.53	305.79	0.1
2	ΔT (℃)	303.35	300.00	300.00	301.12	1.6
3	$T_{1\omega}$(℃)	300.77	297.16	300.62	299.52	2.1
4	R_c (℃/s)	294.24	296.98	299.89	297.04	2.9
5	R_i (℃/s) from linear fitting and uncooled camera	296.02	296.34	285.52	292.62	4.4
*Difference with respect to the fatigue limit from S-N curve (306 MPa) [27]						

Conclusions

A thorough experimental campaign was carried out to examine the effectiveness of five thermal indices in fatigue limit prediction and determine the thermal response of SS 316 samples under cyclic loadings. The use of the same dataset is the strongest point of the comparison. For comparison, the difference between the reference fatigue limit value from the S-N curve and the estimated fatigue limit based on Loung's technique was evaluated.

From the results, $T_{2\omega}$ and R_i from cooled camera performed better than the other thermal indices studied, yielding the closest forecasts to the reference value, with 0.1% difference. ΔT, $T_{1\omega}$ and R_c show greater errors ranging from 1.6% to 2.9%. However, R_i from uncooled camera showed the worst performance, error of 4.4%.

As the main contribution of the work, the results emphasize how time duration for curve fitting and thermal acquisition techniques affect the precision of R_i fatigue limit prediction. The data obtained with a cooled camera yields the most accurate fatigue limit predictions, particularly when the fitting time is more than three seconds. On the other hand, estimations from an uncooled camera exhibit notable oscillation, which makes it difficult to make a reliable fatigue limit estimate. However, if an uncooled camera is the only option, it is recommended a fitting duration of at least 3 seconds for linear fitting to minimize error in fatigue limit estimation.

References

1. J. Schijve, Fatigue of Structures and Materials Springer Dordrecht, 2009.
2. S. Suresh, Fatigue of materials, Cambridge solid state science series, 1992.
3. F. Bjørheim, S.C. Siriwardane, D. Pavlou, A review of fatigue damage detection and measurement techniques, International Journal of Fatigue, 154 (2022) 106556.
4. R. Pippan, C. Zelger, E. Gach, C. Bichler, H. Weinhandl, On the mechanism of fatigue crack propagation in ductile metallic materials, Fatigue & Fracture of Engineering Materials & Structures, 34 (2011) 1-16.
5. G. Antaki, R. Gilada, Chapter 2 - Design Basis Loads and Qualification, in: G. Antaki, R. Gilada (Eds.) Nuclear Power Plant Safety and Mechanical Integrity, Butterworth-Heinemann, Boston, 2015, pp. 27-102.
6. S.M. Shepard, Advances in pulsed thermography, in: Advances in Pulsed Thermography. Proc. SPIE - The International Society for Optical Engineering, Thermosense XXVIII, Orlando, Florida, 2001.
7. C. Ibarra-Castanedo, N.P. Avdelidis, X.P. Maldague, Quantitative pulsed phase thermography applied to steel plates, in: Thermosense XXVII, International Society for Optics and Photonics, 2005.
8. G. Pitarresi, R. Cappello, A. Capraro, V. Pinto, D. Badagliacco, A. Valenza, Frequency modulated thermography-NDT of polymer composites by means of human-controlled heat modulation, in: P. Rizzo, A. Milazzo (Eds.) European Workshop on Structural Health Monitoring, Springer International Publishing, Cham, 2023, pp. 610-618.
9. E. D'Accardi, R. Krankenhagen, A. Ulbricht, M. Pelkner, R. Pohl, D. Palumbo, U. Galietti, Capability to detect and localize typical defects of laser powder bed fusion (L-PBF) process: an experimental investigation with different non-destructive techniques, Progress in Additive Manufacturing, 7 (2022) 1239-1256.
10. G. Dell'Avvocato, D. Gohlke, D. Palumbo, R. Krankenhagen, U. Galietti, Quantitative evaluation of the welded area in resistance projection welded (RPW) thin joints by pulsed laser thermography, SPIE, 2022.
11. N. Montinaro, D. Cerniglia, G. Pitarresi, Evaluation of vertical fatigue cracks by means of flying laser thermography, Journal of Nondestructive Evaluation, 38 (2019) 48.
12. D. Palumbo, R. De Finis, U. Galietti, Thermoelastic stress analysis as a method for the quantitative non-destructive evaluation of bonded CFRP T-joints, NDT & E International, 124 (2021) 102526.
13. M. Zaeimi, R. De Finis, D. Palumbo, U. Galietti, Fatigue limit estimation of metals based on the thermographic methods: A comprehensive review, Fatigue & Fracture of Engineering Materials & Structures, 47 (2024) 611-646.
14. M. Zaeimi, R. De Finis, D. Palumbo, U. Galietti, Numerical Simulation of the Heat Dissipation During the Fatigue Test, in: C. Franck, K. Kasza, J. Estrada, R. De Finis, G. Ólafsson, S. Gururaja, J. Furmanski, A. Forster, P. Kolluru, M. Prime, T. Berfield, C. Aydiner (Eds.) Challenges in Mechanics of Biological Systems and Materials, Thermomechanics and Infrared Imaging, Time Dependent Materials and Residual Stress, Volume 2, Springer Nature Switzerland, Cham, 2024, pp. 83-90.
15. M. Zaeimi, R. De Finis, D. Palumbo, U. Galietti, A Hybrid Approach for Heat Source Identification and Heat Diffusion–Related Issues in the Fatigue Loading, Fatigue & Fracture of Engineering Materials & Structures, n/a (2025).
16. G. La Rosa, A. Risitano, Thermographic methodology for rapid determination of the fatigue limit of materials and mechanical components, International Journal of Fatigue, 22 (2000) 65-73.
17. M. Mehdizadeh, M.M. Khonsari, On the application of fracture fatigue entropy to variable frequency and loading amplitude, Theoretical and Applied Fracture Mechanics, 98 (2018) 30-37.
18. D. Palumbo, U. Galietti, Thermoelastic Phase Analysis (TPA): a new method for fatigue behaviour analysis of steels, Fatigue & Fracture of Engineering Materials & Structures, 40 (2017) 523-534.
19. J.C. Krapez, D. Pacou, G. Gardette, Lock-in thermography and fatigue limit of metals, in: 2000 Quantitative InfraRed Thermography, 2000.
20. R. De Finis, D. Palumbo, U. Galietti, On the relationship between mechanical energy rate and heat dissipated rate during fatigue for a C45 steel depending on stress ratio, Fatigue & Fracture of Engineering Materials & Structures, 44 (2021) 2781-2799.

21. R.D. Finis, D. Palumbo, A. Pirinu, A. Saponaro, F.W. Panella, R. Nobile, U. Galietti, Fatigue behaviour assessment of C45 steel by means of energy-based methods, in: IOP Conference Series: Materials Science and Engineering, 2021.
22. G. Meneghetti, Analysis of the fatigue strength of a stainless steel based on the energy dissipation, International Journal of Fatigue, 29 (2007) 81-94.
23. M. Naderi, M. Amiri, M.M. Khonsari, On the thermodynamic entropy of fatigue fracture, Proceedings of the Royal Society A: Mathematical, Physical and Engineering Sciences, 466 (2010) 423-438.
24. A. Risitano, G. Risitano, Determining fatigue limits with thermal analysis of static traction tests, Fatigue & Fracture of Engineering Materials & Structures, 36 (2013) 631-639.
25. C. Colombo, L. Vergani, Thermographic applications for the rapid estimation of fatigue limit, Procedia Structural Integrity, 24 (2019) 658-666.
26. M. Ricotta, G. Meneghetti, B. Atzori, G. Risitano, A. Risitano, Comparison of experimental thermal methods for the fatigue limit evaluation of a stainless steel, Metals, 9 (2019) 677.
27. A. Mahmoudi, M.M. Khonsari, A Thermodynamic Framework for Rapid Prediction of S-N Curves Using Temperature Rise at Steady-State, Experimental Mechanics, (2023).
28. C. Colombo, M. Sansone, L. Patriarca, L. Vergani, Rapid estimation of fatigue limit for C45 steel by thermography and digital image correlation, The Journal of Strain Analysis for Engineering Design, 56 (2021) 478-491.
29. A. Aghayan, P. Jaiswal, H.R. Siahkoohi, Seismic denoising using the redundant lifting scheme, GEOPHYSICS, 81 (2016) V249-V260.
30. M.P. Luong, Fatigue limit evaluation of metals using an infrared thermographic technique, Mechanics of Materials, 28 (1998) 155-163.

Chapter 6
Investigation of Failure in Unidirectional Adhesively-Bonded Composite Joints via Infrared Thermography

Nithinkumar Manoharan and Suhasini Gururaja

Abstract Adhesive bonding is one of the most effective methods for joining composite materials, preferred over mechanical bonding due to its significant advantages such as reduced weight and uniform stress distribution. Adhesively bonded composite joints are widely used in high cycle fatigue (HCF) applications, including aircraft and wind turbines. While extensive research has been focused on the fatigue performance and failure modes of lap joints due to their ease of manufacturing, it is critical to investigate the fatigue behavior of fully bonded adhesive joints. In lap joints, under cyclic loads, shear and peel stresses are well-studied for their role in damage initiation due to the load path. However, in fully bonded adhesive joints, uniaxial stress across the multi-material system presents various challenges, such as the adhesive-adherend interaction along the longer bondlines. Therefore, understanding the location of damage initiation, the evolution of damage mechanisms, the factors influencing failure, and the failure mode are essential.

This study aims to use infrared thermography (IRT) to evaluate the behavior of fully bonded unidirectional glass fiber-reinforced polymer (UD-GFRP) composite adhesive joints under quasi-static tensile and fatigue loading. The in situ full-field surface temperature variations during cyclic and static loading are recorded using a cooled infrared camera and examined in addition to the displacement field data obtained by two-dimensional digital image correlation (2D-DIC) data to assess damage incipience and progression. Additionally, the recorded surface temperature data is analyzed in the frequency domain to detect second harmonic signals, which are associated with dissipative heat sources and serve as indicators of damage onset and propagation. Pixel-wise analysis of second harmonic signal data provides a visual representation of damage evolution.

Keywords Adhesive bonding · Infrared thermography · Unidirectional glass fiber-reinforced polymer · Digital image correlation · Quasi-static tensile loading

Introduction

In recent decades, composites have become integral to various industries due to their lightweight properties and superior mechanical performance compared to metals. While this adoption continues to grow, the manufacturing of composite materials presents challenges such as high cost, maintenance difficulties, and complex repair processes. Additionally, joining composite structures for assembly and functionality remains a critical manufacturing step. Mechanical joining remains predominant in several industries due to several advantages such as ease of accessibility and repair. However, composite materials face unique challenges with mechanical fastening, such as machining damage, and high stress concentrations at the location of holes drilled for fasteners which could significantly compromise the mechanical performance. Adhesive bonding, on the other hand, requires relatively minimal surface preparation and provides uniform stress distribution while enabling joining of dissimilar materials [1]. As a result of this, huge structures such as wind turbine blades use adhesively bonded interfaces. Despite these advantages, adhesively bonded composite joints are sensitive to manufacturing defects such as trapped voids in the adhesive layer and non-uniform bonding of the adherend-adhesive interface. Such bondline defects could cause localized vulnerable areas that are prone to failure. Delamination and debonding due to the presence of defects have been the leading cause of premature failure in wind turbines after installation [2]

Nithinkumar Manoharan · Suhasini Gururaja
Department of Aerospace Engineering, Auburn University, Auburn, AL 36849
e-mail: nzm0074@auburn.edu; suhasini.gururaja@auburn.edu

Garrett Pataky et al. (eds.), *Fracture, Fatigue, Failure, Damage Evolution and Thermomechanics & Infrared Imaging, Vol. 4*,
Conference Proceedings of the Society for Experimental Mechanics Series, https://doi.org/10.13052/9788743808305-6

Thus, it is vital to understand the damage initiation and progression in such multi-material systems to improve the reliability of adhesive joints. This study aims to understand the behavior of UD-GFRP adhesive joints under quasistatic loading. The full field temperature behavior at the adhesive-adherend interface was observed until the final failure. This was then coupled with full-field strain data using 2D-DIC data to assess the failure mode. To further examine the influence of bondline defects, process defects were introduced during manufacturing for a comparative analysis of variations in temperature and strain in the presence of defects.

Background

IRT is a non-destructive and non-contact technique that has gained a lot of attention in recent decades for inspection and analyzing the damage state of the material [3]. This technique is based on the full field surface temperature measurements to observe its variation with applied excitation, either mechanical or thermal. In isotropic materials, the relationship between stress state and temperature variation in adiabatic and elastic conditions is given by the experimental equation for thermoelasticity,

$$\frac{\Delta T}{T_0} = \frac{-\lambda}{\rho C_p} \Delta\sigma \tag{1}$$

where

T_0 – absolute temperature of the material,
ΔT – change in temperature,
λ – linear thermal expansion coefficient,
ρ – material density,
C_p – specific heat at constant pressure,
$\Delta\sigma$ – change in the sum of principal stresses.

The equation 1 was extended to orthotropic materials, such as polymer composites, by introducing thermoelastic constants in each direction [4]. Under tensile loads, composites show an initial decrease in surface temperature due to the increase in volume in the elastic stage, followed by a gradual increase in temperature with damage onset and propagation. IRT has been used as a technique to visualize damage in composite materials by detecting subtle surface temperature changes, which result from internal damage mechanisms such as matrix cracking, fiber breakage, and delamination. The technique has also been pursued by several authors due to its rapid nature of evaluating the fatigue limit with fewer specimens compared to traditional approaches such as stress life (S-N curve) approach [5]. Passive IRT has been used in this study to detect localized damage progression at the adhesive-adherend interface under static loading.

The DIC technique, developed in the 1980s, has been predominantly used across several industries to measure displacement and strains in a broad range of materials including composites [6]. DIC has also been used to characterize multi-material systems, such as hybrid composite-metal joints [7], demonstrating its applicability to sophisticated engineering problems. To understand the tensile behavior of UD-GFRP joints and to analyze the localized deformation in the adhesive-adherend interface 2D-DIC has been used in this study along with passive IRT.

Material and Methods

Unidirectional glass fiber (UD-GFRP ET 1200-7P) from Vectorply Corporation was used as the adherend system in this study, reinforced with 635 Thin Epoxy and Hardener (3:1) from US Composites. A hot press was used to fabricate the laminates at 80 °C with a pressure of 0.5 tonnes for 30 minutes. The adhesive system consisted of MGS PR 135G3/BPH 1355G resin/hardener adhesive bond paste with a weight ratio of 45:100, mixed using a FlackTek SpeedMixer. This adhesive system is similar to those used in wind turbine blade manufacturing [8]. The laminates were cut to the standard (ASTM D3039) using a WAZER desktop waterjet cutting machine, and an aluminum mold was used for the fabrication of the adhesive joint to ensure precise alignment. The adherends were placed at the bottom mold followed by prepared adhesive bond paste uniformly spread over the adherend using a plastic syringe. Finally, another set of adherends was placed at the top and the mold was tightly closed. The protrusions at the top part of the mold ensured uniform adhesive thickness (2 mm) throughout. The mold was placed in an oven at 75°C for four hours for curing. After curing, the specimens were removed, excess adhesive was polished off, and finally, end tabs were placed. In the case of specimens with process defects, a 20 mm long PTFE Kapton film (25.4 um thick) was fixed to the adherend surface, centered along its length, before applying the adhesive to simulate a bondline defect.

Static tests at 1.3 mm/min loading rate (ASTM D897) were carried out using a Landmark MTS servo-hydraulic system (Model 370.10). A cooled MWIR infrared camera (Telops) with a 50 mm lens, 640x512 pixels resolution and 20 mK NETD was used to record the full-field surface temperature at the adhesive-adherend interfaces. The specimen face from which the specimen temperature was recorded was painted black to ensure homogenous emissivity [9]. In order to perform 2D-DIC analysis, the opposing face was painted white with a random black speckle pattern. Images were captured using a Canon EOS Rebel T7i (18-135 mm lens) and later analyzed using Vic-2D (Correlated Solutions) software.

Results and Discussion

The failure load was 29.76 kN in pristine specimens, while the presence of a bondline defect showed an ~8% reduction in failure load with varying failure mode, not much difference was observed in the failure strain values for both cases. The full field strain field (as microstrains), in addition to the thermograms showing damage initiation and final failure, can be seen in Figure 1. It should be noted that image subtraction was performed using the initial undamaged image as a reference

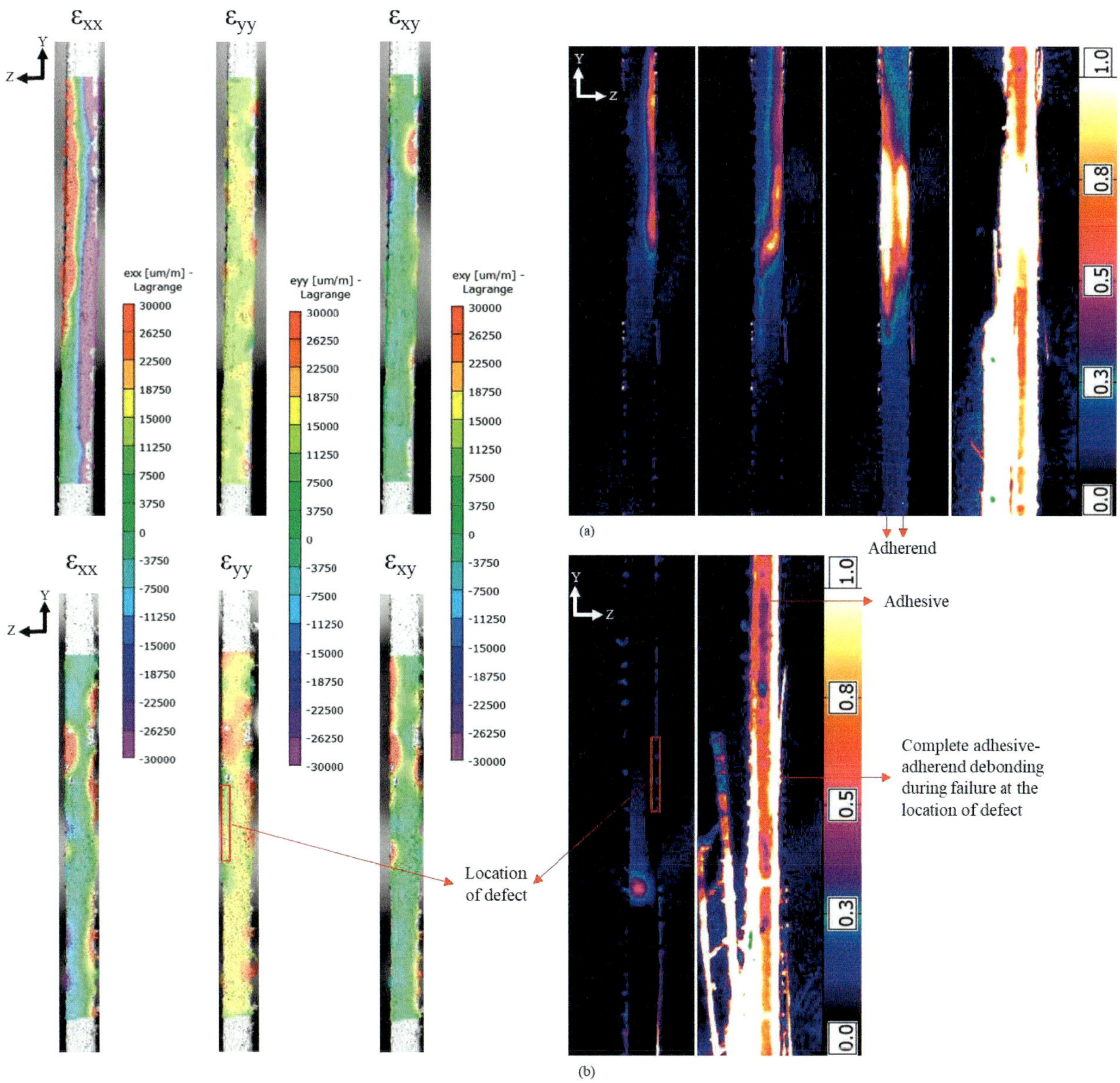

Fig. 1 Full field microstrains before final failure and the thermograms showing change in surface temperature during failure (in °C) after image subtraction with initial state for (a) pristine, and (b) specimen with bondline defect under quasistatic tensile load

to improve the visibility of subtle temperature variations. Pristine specimens and specimens with bondline defect exhibited similar surface temperature behavior throughout the test, where the initial stages of loading showed a decrease in the surface temperature due to the thermoelastic effect, followed by a gradual increase as an indication of damage initiation, and a sudden spike (approximately 20°C in both configurations) during final failure.

The speckle pattern shown clearly distinguishes the adhesive layer from the adherend on either side. The transverse strain (ε_{xx}) field showed a uniform distribution across each layer in pristine specimens, whereas a high strain concentration around the location of defect was seen in the other configuration indicating stress localization. Similar trends were seen when comparing the longitudinal strain (ε_{yy}) contours, where a uniform strain distribution with high values of strain at the location of crack initiation could be seen in pristine specimens, suggesting a more progressive failure, while early stress redistribution around the location of the defect was observed in the defective specimen, leading to a sudden failure [10]. Shear strains (ε_{xy}) showed high values at the interface region where the defect was placed, aligning with a complete interface debond with catastrophic failure at the same location (as seen in the thermograms).

A relatively uniform strain field, with uniform gradual damage accumulation leading to final failure with crack progression from the interface through the adhesive layer causing final failure via crack propagation was observed in pristine specimens. In contrast, specimens with bondline defects showed a non-uniform strain field with localized damage, resulting in sudden, catastrophic failure, highlighting the detrimental effects of even a small defect in the bondline.

Conclusions and Future Work

2D-DIC and surface temperature behavior of adhesively bonded UD-GFRP specimens under quasi-static tensile loads were analyzed. A clear distinction in the full-field strain and temperature behavior was observed, and the combined data provided insight into failure mode in the presence and absence of process defects. The severity of bondline defects was evident, leading to sudden failure and a reduction in final failure load.

Efforts are ongoing to perform cyclic loading on adhesively bonded UD-GFRP specimens. Damage accumulation under fatigue is being studied and the fatigue behavior is evaluated with IRT. The temperature signals throughout the tests are recorded and analyzed in the frequency domain to observe the behavior of the first and the second harmonic components, which have been shown to provide information regarding the damage state of the material [3].

References

1. Borges, C. S. P., et al. "From Fundamental Concepts to Recent Developments in the Adhesive Bonding Technology: A General View." Discover Mechanical Engineering, vol. 2, no. 1, 8, 2023.
2. Mishnaevsky, L. Jr. "Root Causes and Mechanisms of Failure of Wind Turbine Blades: Overview." Materials, vol. 15, no. 9, 2959, 2022.
3. Zaeimi, M., et al. "Fatigue Limit Estimation of Metals Based on the Thermographic Methods: A Comprehensive Review." Fatigue & Fracture of Engineering Materials & Structures, vol. 47, no. 3, 611-646, 2024.
4. Vergani, L., et al. "A Review of Thermographic Techniques for Damage Investigation in Composites." Frattura e Integrità Strutturale, vol. 8, no. 27, 1-12, 2014.
5. Pathak, P., et al. "Examining Infrared Thermography-Based Approaches to Rapid Fatigue Characterization of Additively Manufactured Compression Molded Short Fiber Thermoplastic Composites." Composite Structures, vol. 351, no. 118610, 2025.
6. Janeliukstis, R., et al. "Review of Digital Image Correlation Application to Large-Scale Composite Structure Testing." Composite Structures, vol. 271, no. 114143, 2021.
7. Wang, X., et al. "Characterisation of Composite-Titanium Alloy Hybrid Joints Using Digital Image Correlation." Composite Structures, vol. 140, no. 702-711, 2016.
8. Kuhn, M., et al. "Fatigue Properties of a Structural Rotor Blade Adhesive Under Axial and Torsional Loading." Fatigue & Fracture of Engineering Materials & Structures, vol. 46, no. 3, 1121-1139, 2023.
9. Pothnis, J. R., et al. "Open-Hole Fatigue Testing of UD-GFRP Composite Laminates Containing Aligned CNTs Using Infrared Thermography." Composite Structures, vol. 324, no. 117557, 2023.
10. Chen, J., et al. "Nondestructive Testing and Evaluation Techniques of Defects in Fiber-Reinforced Polymer Composites: A Review." Frontiers in Materials, vol. 9, no. 986645, 2022.

Chapter 7
Effect of Loading Frequency on Fatigue Behavior of Additively Manufactured Short Fiber Thermoplastics via Infrared Thermography

Pharindra Pathak, Kaniz Fatema Bristy, Vipin Kumar, and Suhasini Gururaja

Abstract This study investigates the fatigue behavior of additively manufactured compression-molded short fiber thermoplastics (AM-SFTs), focusing on the effects of loading frequency. Utilizing passive infrared thermography (IRT), the mean temperature changes due to 'self-heating' under cyclic loading applied at varying frequencies were monitored. Previous research demonstrated that IRT could efficiently assess fatigue limits, significantly reducing fatigue characterization time. While traditional methods required up to two weeks and 18 specimens to establish the S-N curve for 20% short carbon fiber-reinforced Acrylonitrile Butadiene Styrene (C/ABS) AM panels, IRT achieved comparable results with just three specimens in a five-hour time frame using a staircase cyclic loading profile [1]. This study builds upon that foundation, investigating the effect of loading frequency on the fatigue behavior of AM-SFTs. In spite of higher loading frequencies generating increased heat due to viscoelastic dissipation and internal friction, which can soften the matrix, the fatigue life of the specimens increased. This heat accumulation, particularly originating at the fiber ends, accelerates inherent progressive damage mechanisms only after crossing the fatigue limit in AM-SFTs. Additionally, microstructure - fiber length distribution, fiber orientation distribution, and porosity distributions - play a significant role in determining the structural performance of AM-SFTs. Micro-computed tomography characterization of AM-SFTs provided information about the composite microstructure. Fatigue testing of AM-SFTs and baseline ABS were performed at varying loading frequencies of 5 Hz and 20 Hz using a servo-hydraulic machine to delineate the role of fibers on the self-heating phenomenon. Additionally, dynamic mechanical analysis (DMA) testing was conducted on AM-SFTs and baseline ABS to assess the effect of fibers on viscoelastic behavior.

Keywords AM-SFTs · IRT · Fatigue · Frequency · Damage

Introduction

Short fiber-reinforced thermoplastic composites (SFTs) have garnered significant attention for structural applications due to their lightweight nature, cost-effectiveness, and ease of manufacturing. Their superior fatigue resistance makes them particularly valuable in the automotive and aerospace industries. Despite its advantages, the fatigue behavior of SFTs is complex and significantly affected by factors such as fiber orientation, porosity, stress ratio, and loading frequency [2]. Under cyclic loading, SFTs typically exhibit continuous softening, characterized by a gradual reduction in stiffness and load-bearing capacity. This behavior is primarily attributed to the initiation and progression of damage within the matrix, as well as at fiber ends and the fiber-matrix interface [3]. Due to the viscoelastic nature of thermoplastics and their relatively low melting temperatures, loading frequency significantly influences fatigue performance, primarily through heat buildup and time-dependent deformation [4]. The fatigue resistance of SFTs is also governed by factors such as fiber reinforcement, orientation, and distribution, which are inherently linked to the effective material response of the SFT [5]. Due to the complexity and the numerous parameters affecting the mechanical behavior of SFTs, fatigue behavior has primarily been studied

Pharindra Pathak · Kaniz Fatema Bristy · Suhasini Gururaja
Department of Aerospace Engineering, Auburn University, Auburn, AL 36849
e-mail: pzp0057@auburn.edu; kzb0153@auburn.edu; szg0130@auburn.edu

Vipin Kumar
Manufacturing Science Division, Oak Ridge National Laboratory, Oak Ridge, TN 37830
e-mail: kumarvi@ornl.gov

Garrett Pataky et al. (eds.), *Fracture, Fatigue, Failure, Damage Evolution and Thermomechanics & Infrared Imaging, Vol. 4*,
Conference Proceedings of the Society for Experimental Mechanics Series, https://doi.org/10.13052/9788743808305-7

through experimental methods. Traditional fatigue assessment methods, such as the stress-life (S-N) approach, require extensive experimental testing, rendering them both time- and resource-intensive. This highlights the necessity for dependable, efficient methods to predict fatigue life and comprehend damage mechanisms in SFTs.

Infrared thermography (IRT) has emerged as a promising non-destructive evaluation (NDE) tool for real-time monitoring of damage evolution, leveraging heat dissipation during cyclic loading [1, 6, 7, 8]. By correlating dissipated hysteretic energy with surface temperature changes, IRT enables rapid fatigue characterization. Under constant amplitude fatigue (CAF), the surface temperature evolution typically exhibits three distinct phases: an initial rapid temperature rise (Phase I), followed by a stabilized mean temperature indicating steady-state damage (Phase II), and finally, an unsustainable rapid temperature increase (Phase III) that precedes failure. Notably, the stabilized Phase II temperature (ΔT_{stab}) increases with the applied cyclic stress amplitude, and plotting ΔT_{stab} against the stress amplitude yields a bilinear curve that defines the fatigue limit. Furthermore, by employing a staircase-loading protocol where blocks of increasing CAF amplitudes are applied to a single specimen, the fatigue limit of a material system can be determined with minimal specimen usage. Previous studies have demonstrated that the IRT method can accurately capture the fatigue limit for 20 wt.%C/ABS SFTs manufactured via an additive manufacturing compression molding (AM-CM) process, achieving a 100-fold increase in testing speed while requiring only one-sixth the number of specimens compared to traditional approaches [1]. This efficiency of the IRT method instills confidence in its potential for the study of SFTs, particularly those processed via advanced techniques like AM-CM.

In this study, the fatigue behavior of 20 wt.% C/ABS AM-SFTs was investigated. By leveraging IRT, a rapid and efficient method for predicting fatigue limit is established. The influence of loading frequency and fiber orientation on the fatigue limit of 20 wt.% C/ABS AM-SFTs have been investigated in the current study.

Materials and Methods

Materials and Characterization

This study investigated two material systems: pure ABS and ABS reinforced with 20 wt.% chopped C-fibers. These materials were manufactured at Oakridge National Laboratory using a two-step process. For the composite material, chopped carbon fibers were first mixed with ABS pellets and then extruded using a KUKA robotic arm operating at 1200 rpm, with a nozzle speed of 280 mm/s, and a 9.88 mm gap between the substrate and the nozzle outlet [9]. This process created an AM preform, which underwent compression molding at a pressure of 500 kN for 5 minutes. The final product results from the combination of the AM process promoting high fiber alignment and the CM process, which reduces porosity, ultimately enhancing the composite's mechanical performance [9].

The AM-CM process, a key aspect of this study, creates a unique microstructure. The chopped fibers tend to align along the bead (deposition) direction, resulting in directionally dependent properties. This unique microstructure provides a novel platform to assess the effect of material orientation on mechanical properties and fatigue behavior and to compare these results with the pure ABS matrix. Specimens were machined at different angles relative to the bead direction (0°, 30°, 45°, 60°, and 90°) using abrasive waterjet machining following ASTM D638 Type I specimen dimensions as seen in Fig. 1(a).

To further investigate the microstructure, a specimen along bead direction (0°) was subjected to high-resolution 3D X-ray imaging using a Zeiss Xradia 620 Versa system at Auburn University. For optimal image resolution during μ-CT acquisition, each specimen was mounted on a rotary stage perpendicular to the X-ray source. μ-CT data were collected at a power setting of 140 kV and 21 W using a 4× scintillator objective lens with 2×2 binning, resulting in a voxel size of approximately 0.6 μm × 0.6 μm × 0.6 μm. An air filter was used to enhance image clarity. After acquiring X-ray projection images, a 3D reconstruction was performed using the system's image processing unit, generating 997 images stacked in the out-of-plane direction. Noise reduction and intensity-based thresholding were applied to determine the fiber, which is seen as red in color (Fig. 1(b)), pores, and matrix constituents. 0ptimized AM-CM processing yield up to 80% fiber alignment in the deposition direction, within the 0°–30° range, with a typical fiber aspect ratio of 5–7 and an average fiber diameter of 8 μm as shown in Fig. 1(c).

Quasi-Static Tensile Testing

Quasi-static tensile tests were conducted on pure ABS and composite specimens with varying fiber orientations using a 100 kN servo-hydraulic MTS machine at a displacement rate of 2 mm/s. For each specimen type, three replicates were tested to ensure statistical reliability. The measured mechanical properties of ultimate tensile strength (σ_{UTS}) and failure strain (ε_f) are summarized in Table 1.

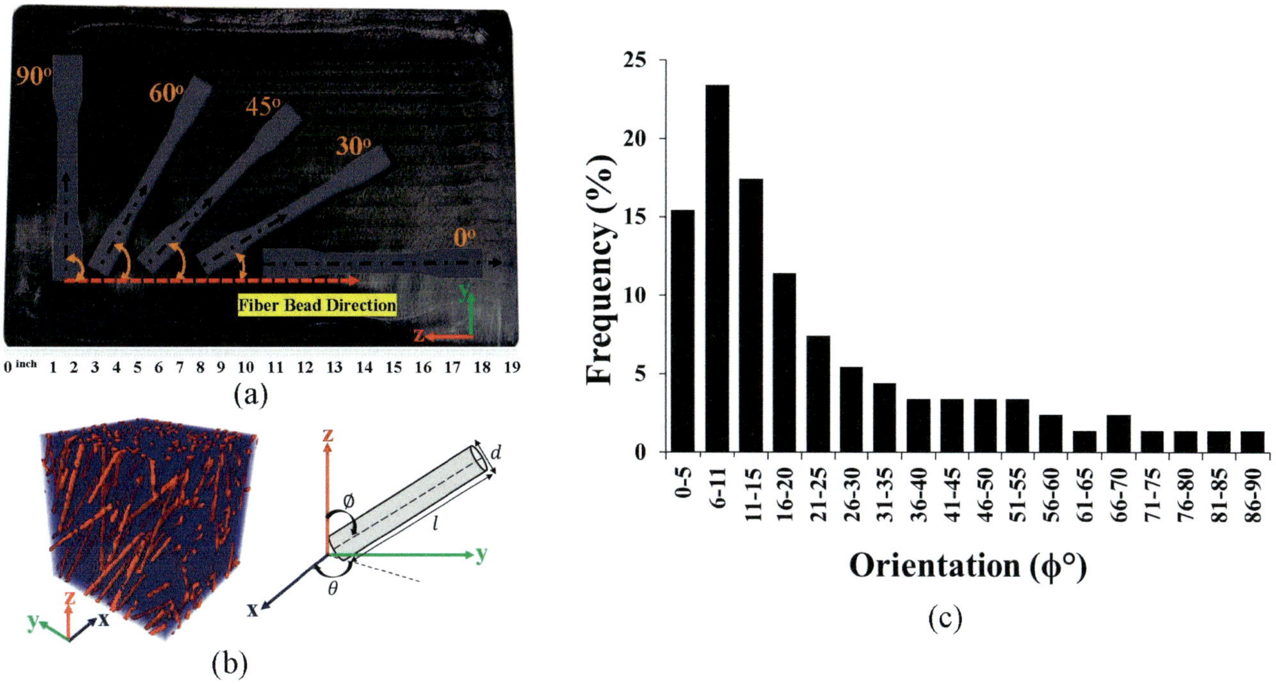

Fig. 1 (a) Schematic of AM-CM composite panel with fiber bead direction and different oriented specimens; (b) Visualization of fibers from μ-CT, and semi-axes and orientation conventions for a geometric representation of fiber. The aspect ratio is given by l/d, and (c) Distribution of fiber parameters obtained from μ-CT scan of orientation ϕ.

Table 1 Ultimate tensile strength and failure strain for various specimens.

Specimen	Ultimate Tensile Strength, σ_{UTS} [MPa]	Failure Strain, ε_f [%]
Pure ABS	54.000 ± 1.810	1.320 ± 0.050
0° fiber orientation	87.320 ± 1.240	0.730 ± 0.003
30° fiber orientation	61.720 ± 0.770	0.910 ± 0.001
45° fiber orientation	44.600 ± 0.190	0.935 ± 0.005
60° fiber orientation	43.320 ± 1.490	0.877 ± 0.002
90° fiber orientation	42.420 ± 1.180	1.055 ± 0.004

The results indicate a strong dependence of σ_{UTS} and ε_f on fiber orientation. Specimens with fibers aligned at 0° exhibited the highest σ_{UTS} and the lowest ε_f, as expected. At 0°, the fibers carry most of the applied load efficiently making the material stiffer but brittle, resulting in restriction in matrix deformation. As the fiber orientation deviates from 0°, σ_{UTS} decreases while ε_f increases, indicating that transverse fiber orientations (e.g., 90°) provide minimal load-bearing capacity and are more matrix-dominated. Interestingly, for fiber orientations 45° to 90°, σ_{UTS} was found to be lower than pure ABS matrix. These changes in behavior for different orientations are primarily due to the load transfer efficiency between the fibers and the polymer matrix, which depends on fiber alignment relative to the loading direction.

Fatigue Testing

To evaluate the fatigue behavior under cyclic loading, interrupted fatigue tests were performed using the same 100 kN MTS machine Fig. 2(a). The tests were carried out at two frequencies: 5 Hz and 20 Hz under a constant stress ratio ($R = 0.1$). Each specimen was initially loaded at a stress level corresponding to 30% of its σ_{UTS}, and the stress was incrementally increased by 5% until failure. Between each stress increment, the specimen was cycled for 7500 cycles, with a rest period provided to allow the specimen to return to ambient temperature Fig. 2(b). This protocol enabled the specimen to experience a range of stress levels and capture various damage modes.

Surface temperature evolution due to cyclic loading-induced self-heating was monitored using a Telops infrared camera Fig. 2(a). The camera recorded the surface temperature increment (Fig.3(a))and generated temperature profiles at every 7500-cycle interval Fig. 3(c). The 7500-cycle step size was based on constant amplitude fatigue tests, demonstrating that

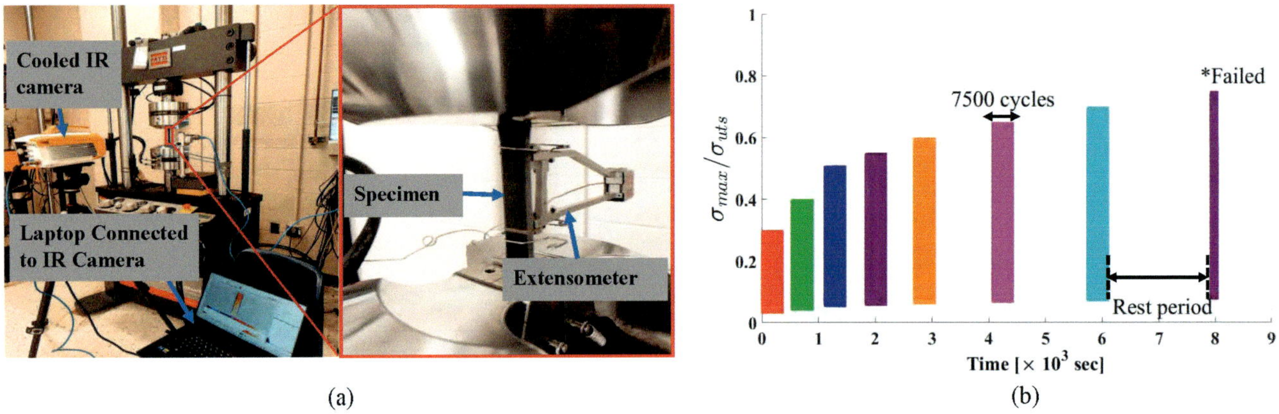

Fig. 2 (a) Experimental setup for in situ surface temperature measurements, featuring a cooled IR camera connected to a laptop with Reveal IR software, integrated with a UTM system and a 5 cm gage length extensometer, and (b) Staircase loading applied to a specimen to determine fatigue limit rapidly: early increments of 10% σ_{UTS} followed by 5% σ_{UTS} increments to $\sigma_{\max}$ were used to determine fatigue limit. The stress ratio $\mathrm{R} = 0.1$ was maintained throughout the test. The rest period between load steps ensured the specimen cooled to room temperature.

the specimen's temperature stabilized after approximately 6000 cycles, even at higher stress levels. Initially, as the specimen experienced cyclic loading, the temperature increased rapidly (Phase I) due to the viscoelastic response of the ABS matrix and friction at the fiber-matrix interface, particularly at the fiber ends where stress concentrations occur. This rapid increase was followed by a plateau (Phase II), representing a saturation of damage accumulation. At higher stress levels, the energy input exceeded the bonding strength at the fiber-matrix interface, leading to matrix cracking, increased displacement, and fiber-matrix debonding. These phenomena culminated in a final rapid temperature spike (Phase III) immediately preceding failure.

Fast Fourier transform (FFT) analysis of the temperature data as explained in [6] for composite specimens tested at 20 Hz revealed that inelastic or irreversible damage initiates at stress levels above 60% of σ_{UTS}, with failure occurring near 75% of σ_{UTS} Fig. 3(b). The temperature profiles clearly indicate that higher stress levels result in more significant temperature increments beyond the fatigue onset point.

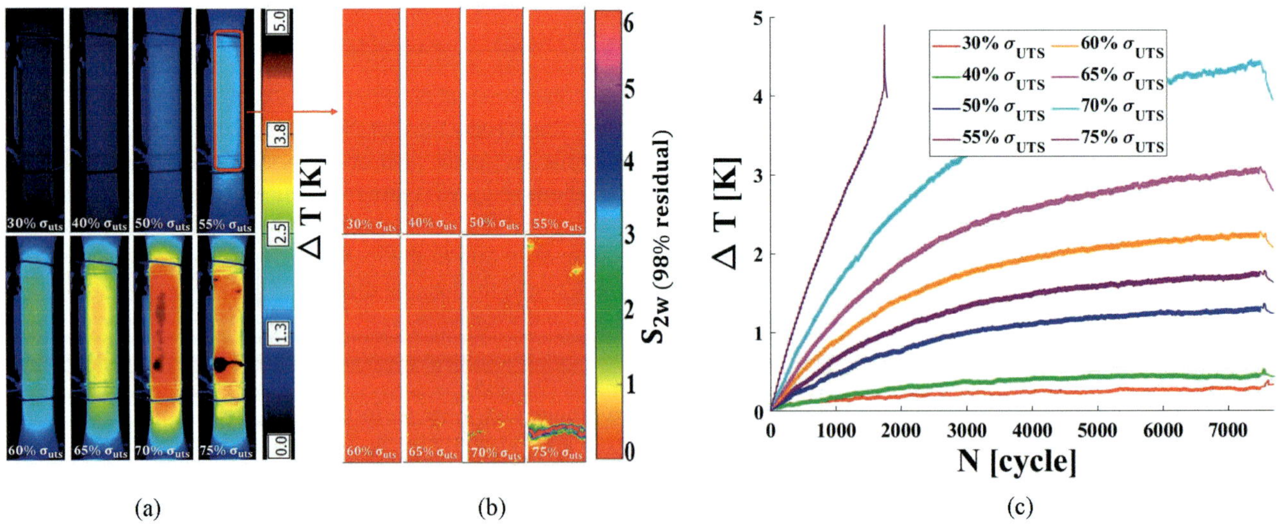

Fig. 3 (a) Stabilized full-field surface temperature profile of a 20 wt.% C/ABS specimen at 20 Hz load frequency with increasing load steps indicated in each sub-figure, (b) Evolution of irreversible damage within gage section using second harmonics approach, and (c) Evolution of maximum surface temperature of the specimen under staircase loading. The specimen was allowed to cool to room temperature after each load step.

Results and Discussion

Comparison of Pure ABS with C/ABS

In this study, the temperature rise observed in pure ABS and C/ABS composites under cyclic loading was compared to understand the contributions of fiber-matrix interactions versus the behavior of the pure matrix. For pure ABS, a significant temperature increase was observed under cyclic loading at both 5 Hz and 20 Hz. Since ABS is a viscoelastic polymer, DMA and differential scanning calorimetry (DSC) were performed to confirm a glass transition temperature of approximately 105 °C. At room temperature (25 °C), the primary source of energy dissipation in pure ABS during fatigue loading arises from internal molecular friction, mechanical damage, and creep effects rather than thermal degradation. At higher frequencies (20 Hz), the molecular chains are forced to respond to the applied load more rapidly, leading to higher levels of internal friction. As a result, energy dissipation increases at 20 Hz, which manifests as a higher temperature rise in pure ABS (see Fig. 4(a)-(b)).

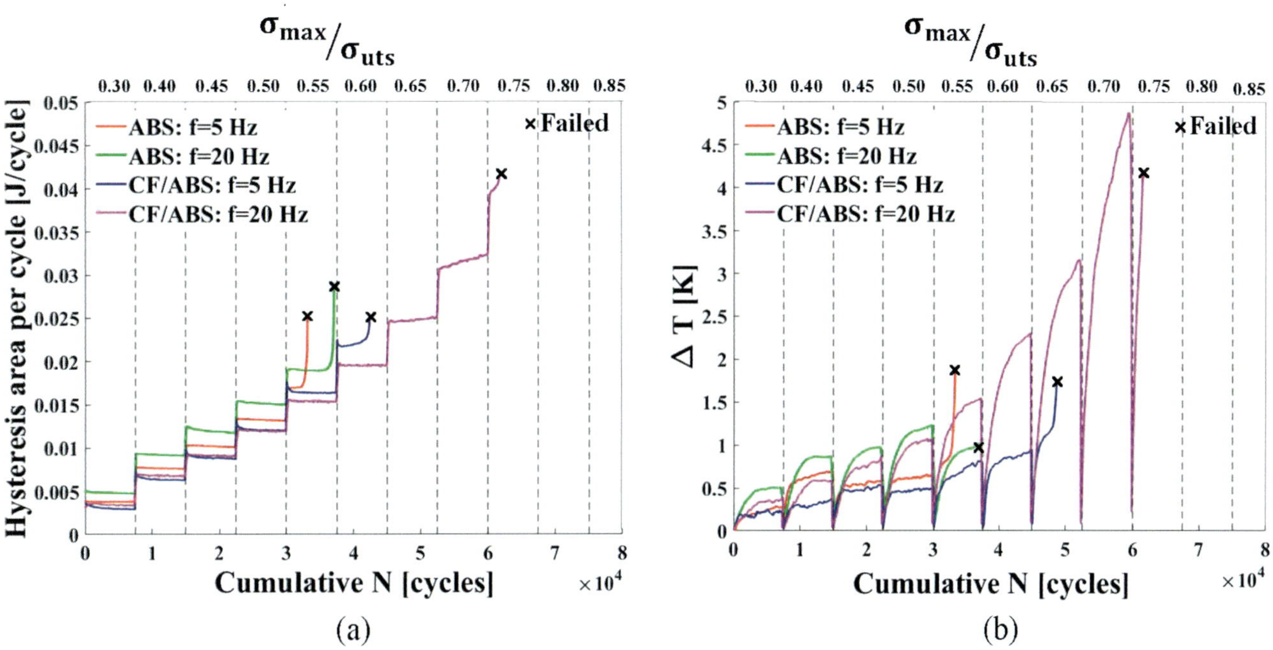

Fig. 4 Comparison of pure ABS and C/ABS composite for two different frequencies of 5 Hz and 20 Hz: (a) hysteresis energy loss per cycle, and (b) typical surface temperature evolution with cycle.

For the C/ABS composite, the fiber-matrix interaction played a significant role in temperature rise, especially at higher frequencies. At 5 Hz, the loading time was long enough for the polymer matrix to deform more significantly, with the carbon fibers restricting the mobility of the polymer chains. This resulted in increased internal friction and greater energy dissipation within the matrix, which contributed to a higher temperature rise. Furthermore, frictional interactions at the fiber-matrix interface added to the energy dissipation, especially at the fiber ends and the unsized fiber caps, which acted as stress concentrators. These micro-slip effects at the interface contributed to localized heating in the composite, resulting in a more pronounced temperature increase than pure ABS once the fatigue onset happened. This suggests that, at low frequencies, the composite's matrix behavior is still crucial in determining energy dissipation, with the fibers acting to restrict the matrix's deformation and enhance internal friction.

At higher frequencies (20 Hz), the fibers in the composite carried a greater portion of the load, reducing the strain on the polymer matrix. As a result, the energy dissipation from polymer chain movement was lower (see Fig. 4(a)), and the temperature rise was less influenced by the polymer matrix's viscoelastic behavior. However, the friction at the fiber-matrix interface became more pronounced due to localized micro-slip and fiber-matrix interactions, leading to additional heat generation. The thermal conductivity of the composite also helped in distributing heat more efficiently at lower stress levels, allowing for better overall heat distribution. Yet, at higher stress levels, fiber-matrix interfaces became active sites for micro-damage, which contributed to localized heating that exceeded the temperature rise observed in pure ABS.

On comparing both pure ABS and C/ABS, in pure ABS, the dominant mechanism of energy dissipation is molecular friction within the polymer matrix. At higher frequencies, the chains cannot fully rearrange and respond, causing more friction and, consequently, more heat. For composites, at lower stress levels, carbon fibers enhance the thermal conductivity of the composite, meaning heat generated from friction and energy dissipation can spread more quickly within the material, allowing for better heat distribution. However, at higher stress levels, fiber ends and fiber-matrix interfaces become active sites for micro-damage and frictional interactions, contributing to localized heating. This results in a temperature increase that can exceed that of pure ABS, as fatigue damage accumulates over cycles (see Fig. 4(b)).

Effect of Loading Frequency on the Fatigue Limit

The influence of loading frequency on the fatigue behavior of the C/ABS composite was investigated by subjecting specimens to interrupted fatigue tests at 5 Hz and 20 Hz. In pure ABS, displacement remained almost constant at both frequencies; however, the hysteresis loss was more significant at 20 Hz, leading to a higher temperature increase (see Fig. 4(a)). In contrast, the C/ABS composite exhibited markedly different displacement behavior. At lower stress levels, the composite experienced higher displacement at 5 Hz due to the longer cycle duration, which allowed the ABS matrix more time to relax. At 20 Hz, the shorter cycle time restricted matrix relaxation, reducing displacement (see Fig. 5(a)).

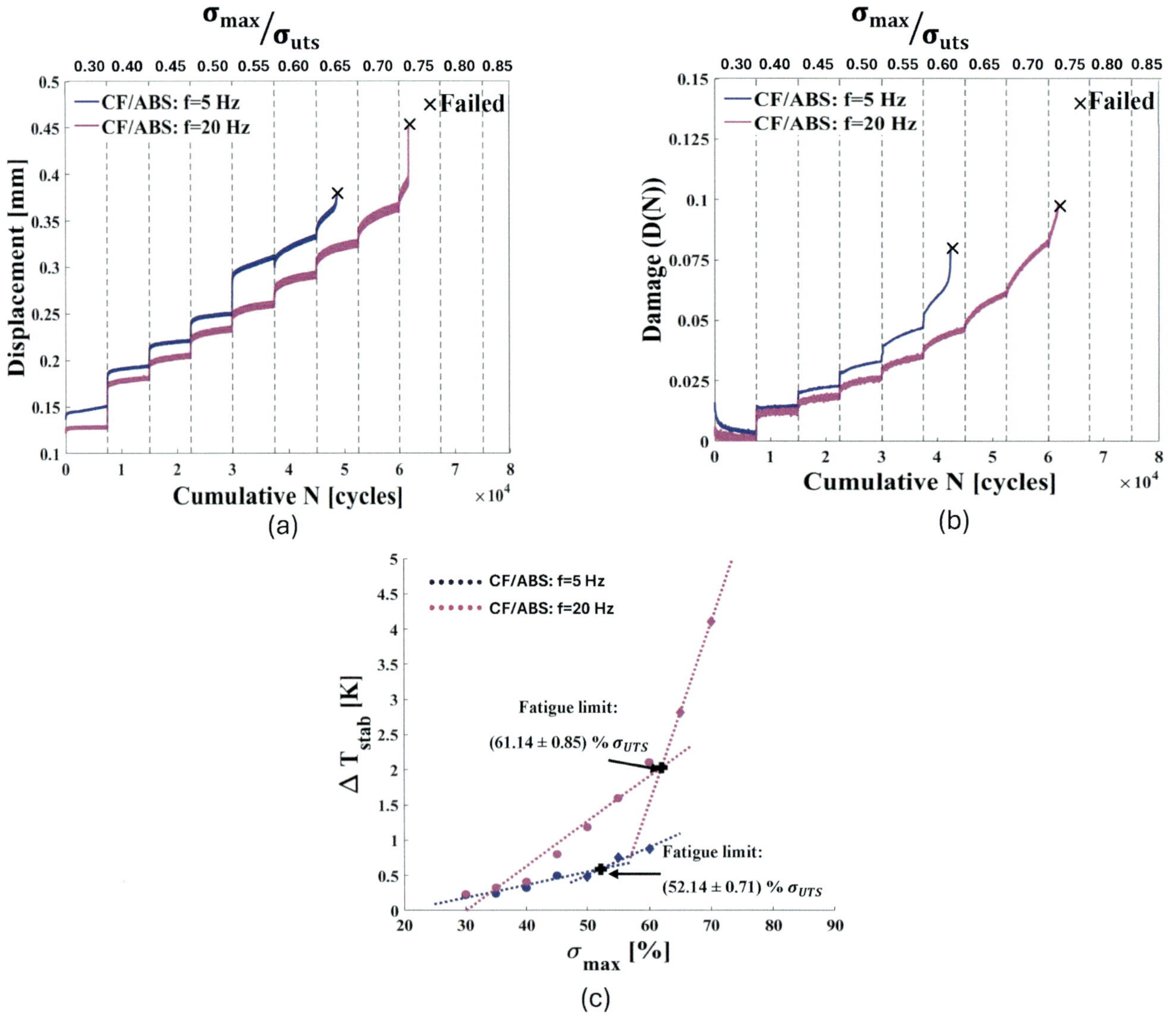

Fig. 5 Comparison of C/ABS composite for two different frequencies of 5 Hz and 20 Hz: (a) displacement for given stress level for given cycles, (b) damage progression, and (c) fatigue limit estimation of given specimens.

As stress levels increased, the displacement at 5 Hz became significantly higher than at 20 Hz. The slower loading cycles at 5 Hz promoted more extensive creep and energy dissipation, leading to higher hysteresis loss and more pronounced damage progression as seen in Fig. 5(b). In contrast, at 20 Hz, the reduced time for thermal buildup resulted in lower displacement and delayed onset of damage. Using a bilinear approach based on the stabilized temperature data, the fatigue limit of the composite was determined to be 52% of its ultimate tensile strength (σ_{UTS}) at 5 Hz and 61% σ_{UTS} at 20 Hz (Fig. 5(c)). These results indicate that the C/ABS composite can withstand more of its σ_{UTS} at increased loading frequencies, likely due to reduced damage accumulation.

Effect of Fiber Orientation on the Fatigue Limit

Fiber orientation relative to the loading direction significantly influences the fatigue performance of the C/ABS composite. All these different fiber-oriented specimens undergo interrupted fatigue loading at 20 Hz frequency with a stress ratio of 0.1. In fatigue tests, 0° specimens experienced longer fatigue lives despite exhibiting higher temperature spikes (see Fig. 6(a)-(b)). The increased temperatures result from concentrated self-heating at the fiber ends, which are stress concentrators. As the orientation of the fiber deviated from 0° with specimens oriented at 30°, 45°, 60°, and 90°, the mechanical response became more dominated by the matrix. Specimens with orientations of 45°, 60°, and 90° not only demonstrated lower σ_{UTS} (comparable to or lower than pure ABS) but also showed earlier failure in fatigue testing (see Table 2).

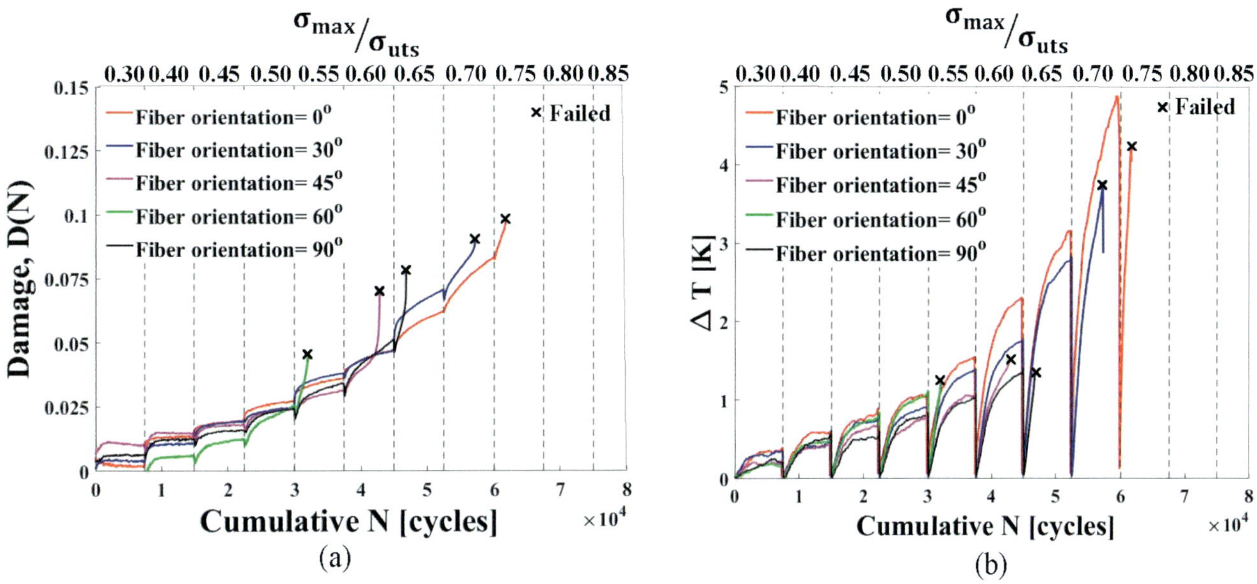

Fig. 6 Comparison of different fiber-oriented C/ABS composite at 20 Hz: (a) damage progression, and (b) typical surface temperature evolution with cycle.

Table 2 Fatigue limit for different configurations.

Specimen	Fatigue limit, [% of σ_{UTS}]
Pure ABS	51.850 ± 0.740
0° fiber orientation	61.950 ± 0.850
30° fiber orientation	56.040 ± 1.080
45° fiber orientation	50.820 ± 0.710
60° fiber orientation	36.180 ± 0.250
90° fiber orientation	38.810 ± 0.780

The damage evolution during cyclic loading was strongly dependent on fiber orientation. The 0° specimens, with fibers oriented along the load direction, exhibited the slowest damage accumulation due to efficient load transfer and effective crack bridging. In contrast, off-axis specimens (e.g., 90°) experienced rapid damage progression, as fibers could not efficiently share the load, leading to increased matrix cracking and fiber-matrix debonding. Temperature profiles further supported these findings: while 0° specimens displayed the highest temperature rise due to more extensive energy dissipation, this did not compromise their superior fatigue performance. As expected, the lower temperature rise observed in the 90 ° specimens was linked to a decreased fatigue limit.

Conclusions and Future Work

The study demonstrates that loading frequency significantly influences the fatigue behavior of the C/ABS composite. Increasing the loading frequency enhances the fatigue limit, primarily due to reduced creep effects and diminished hysteresis energy loss per cycle. While the composite exhibits a lower temperature rise than pure ABS at low-stress levels, at higher stress levels, the composite shows a pronounced temperature increase up to an additional 5 K attributed to fiber–matrix interactions and localized heating.

Furthermore, fiber orientation relative to the bead direction strongly affects the mechanical properties. Specimens with off-axis orientations (45°, 60°, and 90°) behave more like the pure ABS matrix, exhibiting lower ultimate tensile strength and higher failure strain, whereas 0° oriented specimens benefit from efficient load transfer and superior fatigue resistance.

Future work will investigate the effect of stress ratio on fatigue behavior and develop an empirical model for fatigue damage accumulation based on the experimental data.

References

1. Pathak P., Gururaja S., Kumar V., Nuttall D., Mahmoudi A.and Khonsari M.M., and Vaidya U. Examining infrared thermography based approaches to rapid fatigue characterization of additively manufactured compression molded short fiber thermoplastic composites. *Composite Structures*, 351, 2025.
2. Seyyedvahid Mortazavian and Ali Fatemi. Fatigue of short fiber thermoplastic composites: A review of recent experimental results and analysis. *International Journal of Fatigue*, 102:171-183, 2017.
3. Wang S.S. and Chim ES.M. Fatigue damage and degradation in random short-fiber smc composite. *Journal of Composite Materials*, 17:114-134, 1983.
4. Bernasconi A. and Robb M.K. Effect of frequency upon fatigue strength of a short glass fiber reinforced polyamide 6: A superposition method based on cyclic creep parameters. *Polymer Composites*, 30:154-61, 2009.
5. Seyyedvahid Mortazavian and Ali Fatemi. Fatigue behavior and modeling of short fiber reinforced polymer composites including anisotropy and temperature effects. *International Journal of Fatigue*, 77:12-27, 2015.
6. Rosa GL and Risitano A. Thermographic methodology for rapid determination of the fatigue limit of materials and mechanical components. *International Journal of Fatigue*, 2000.
7. Khonsari M and Amiri M. Introduction to thermodynamics of mechanical fatigue. *CRC Press*, 2013.
8. Manoharan N, Pathak P, Gururaja S, Kumar V, and Vaidya U. Rapid fatigue characterization via infrared thermography of am-cm composites. In Springer Nature Switzerland, editor, *Challenges in Mechanics of Biological Systems and Materials, Thermomechanics and Infrared Imaging, Time Dependent Materials and Residual Stress*, volume 2, pages 53-58. 2023.
9. Kumar V, Alwekar SP, Kunc V, Cakmak E, Kishore V, Smith T, Lindahl J, Vaidya U, Blue C, Theodore M, Kim S, and Hassen AA. High-performance molded composites using additively manufactured preforms with controlled fiber and pore morphology. *Additive Manufacturing*, 37:101733, 2021.

Chapter 8
Analytical Generalization of the T and A-theta Integrals for the Study of Cracking in Three-dimensional Orthotropic Medium

Loïc Chrislin Nguedjio, Rostand Moutou Pitti, Benoit Blaysat, Pierre Kisito Talla, Fréderic Dubois, and Naman Recho

Abstract An analysis of the fracture phenomenon in wood shows that it is significantly influenced not only by mechanical loads and the heterogeneous nature of this material but also by multiple parameters related to thermal and moisture variations. It becomes important to propose predictive models that account for these climatic parameters in order to better assess the risks of failure and ensure the safety of wooden structures. In this paper, a new mathematical formulation of the invariant integral, **T** and **A**-theta is proposed to evaluate the energy release rate during the determining od fracture toughness. The formulation is extended to the three-dimensional case and is based on a generalization of Noether's theorem as a free energy that includes both real and virtual fields for mechanical and moisture loads. The resulting model allows for a decoupling of mixed fracture modes and the analysis of the impact of orthotropy and wood shrinkage-swelling on cracking parameters. The independence of the integration contour is then verified, and a finite element procedure is implemented using the Cast3M software to validate the proposed model.

Keywords Fracture · Path-independent integral · Moisture fields · Orthotropic materials · Three-dimensional modelling

Introduction

In the diagnosis and maintenance of civil engineering structures, crack detection is a crucial aspect to ensure the safety of both the structures and their users. This scientific field presents several challenges, particularly for wooden structures, due to the heterogeneous and anisotropic nature of wood, whose overall behavior is significantly influenced by variations in humidity and temperature [1]. Fracture mechanics provides a framework for analyzing the behavior of materials containing cracks and for determining the conditions necessary for crack initiation, propagation, and arrest. It is generally based on two approaches: the local approach and the global approach. The local approach relies on the formulation of continuous stress and displacement fields, which are disturbed by singularities at the crack tip [2]. The global or energetic approach is based on the energy balance of the structure, enabling the isolation of the energy release rate required for the creation of new crack surfaces. To quantify this energy while overcoming singularity issues, invariant integrals appear to be a promising alternative under certain assumptions, as demonstrated by numerous studies in the literature [3, 4].

Loïc Chrislin Nguedjio · Rostand Moutou Pitti · Benoit Blaysat · Naman Recho
Université Clermont Auvergne, Clermont Auvergne INP, CNRS, Institut Pascal, F-63000, Clermont-Ferrand, France
e-mail: loic.nguedjio@doctorant.uca.fr; rostand.moutou_pitti@uca.fr; Benoit.BLAYSAT@uca.fr; naman.recho@uca.fr

Loïc Chrislin Nguedjio · Pierre Kisito Talla
Unité de Recherche de Mécanique et de Modélisation des Systèmes des Physiques (UR2MSP) ; Département de Physiques, Faculté des Sciences, Université de Dschang, BP 67, Dschang, Cameroun
e-mail: loic.nguedjio@doctorant.uca.fr; tpierrekisito@yahoo.com

Rostand Moutou Pitti
CENAREST, IRT, BP14070, Libreville, Gabon
e-mail: rostand.moutou_pitti@uca.fr

Fréderic Dubois
University of Limoges, GC2D, UR 14477, 19300, Egletons, France
e-mail: frederic.dubois@unilim.fr

Garrett Pataky et al. (eds.), *Fracture, Fatigue, Failure, Damage Evolution and Thermomechanics & Infrared Imaging, Vol. 4*,
Conference Proceedings of the Society for Experimental Mechanics Series, https://doi.org/10.13052/9788743808305-8

Among the most commonly used invariant integrals, Rice's **J** integral [5] allows for the calculation of the energy release rate in an elastic cracked medium under pure opening or shear fracture modes. However, in the case of wood, which is often subject to mixed-mode cracking, this integral presents limitations, necessitating the use of another integral, namely the **M** integral [6]. Moreover, these classical integrals are restricted to two-dimensional problems and do not take into account the three-dimensional effects, which are essential when dealing with an orthotropic material such as wood [7]. Beyond the consideration of three-dimensional effects, it is also necessary to extend the modeling approach to variable environments [1]. The first studies on crack modeling in wood using invariant integrals in variable climatic conditions date back to the work of Moutou Pitti et al. [8], further developed by Hassen et al. [1] and Seif et al. [9] in their study of temperature effects through the **A** integral.

This paper focuses on modeling the effect of humidity and incorporating a moisture gradient into the analytical formulation of the **A** integral. The mathematical analyses are revisited, leading to a new expression of the **A** integral, first in 2D and then extended to 3D. A numerical implementation scheme is proposed, along with results illustrating the influence of humidity on the energy release rate.

Background

The formulation of the **T** integral is derived from Noether's theorem [10] and a combination of real and virtual mechanical fields, extended to kinematically admissible moisture content variations. The system's free energy F* is then expressed in a form that incorporates real and virtual displacement fields, respectively $(u,\ v)$, real stress fields ${\sigma_{ij}}^u$ and virtual stress fields ${\sigma_{ij}}^v$, as well as the moisture content field H^u, considered only in a real configuration, all of which remain kinematically admissible. By applying Noether's theorem and imposing the total Lagrangian null condition $\left(\delta L = \int_V \int_t \delta F^* dV dt = 0\right)$, successive integrations lead to the following expression for the **A** integral, established to the same steps as Moutou Pitti et al. [8]:

$$A\theta^{3D} = -\frac{1}{2}\int_V \left[\sigma_{ij,k}^{(v)} \cdot u_i - \sigma_{ij}^{(u)} v_{i,k} - H_{,k}^{(u)} \cdot v_i + H^{(u)} \cdot v_{i,k}\right] \cdot \theta_{k,j} \cdot dV - \frac{1}{2}\int_{S_{CF_+}} \left[\left(\sigma_{ij}^{(v)} u_{i,k} + \left(\sigma_{ij}^{(u)} - H^{(u)}\right) v_{i,k}\right)\right] \theta_k \cdot dS$$
$$+ \frac{1}{2}\int_V \left[\left(\sigma_{ij}^{(v)} u_{i,k}\right)_{,j} + \left(\left(\sigma_{ij}^{(u)} - H^{(u)}\right) v_{i,k}\right)_{,j} - \sigma_{ij}^{(v)} \left(u_{i,j}\right)_{,k} - \left(\sigma_{ij}^{(u)} - H^{(u)}\right) \left(v_{i,j}\right)_{,k} + H_{,k}^{(u)} v_{i,j} - H_{,j}^{(u)} \cdot v_{i,k}\right] \theta_k \cdot dV \quad (1)$$

θ is a field with specific properties designed to circumvent boundary issues in finite element calculations. It is equal to one within the integration contour V and zero in the external domain while remaining continuously differentiable within V (Figure 1). The integral is divided into three main terms, with an extension to steady-state behavior represented by the highlighted quantity. The first term corresponds to the energy required for stationary crack growth while accounting for hygroscopic effects and is used for mixed-mode decoupling. The second term introduces a pressure effect on the crack lips, while the third term facilitates the separation of mixed modes during crack propagation.

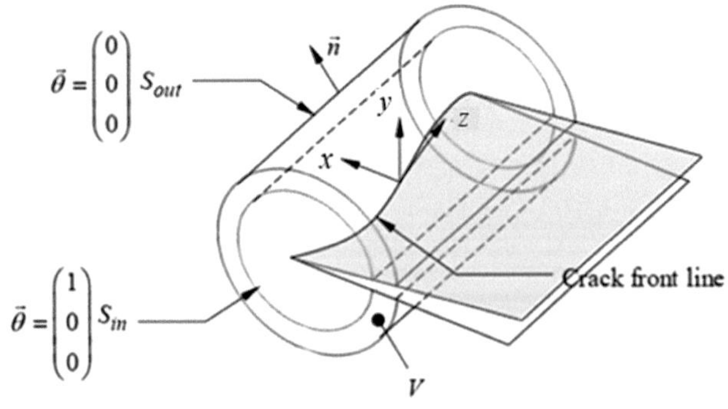

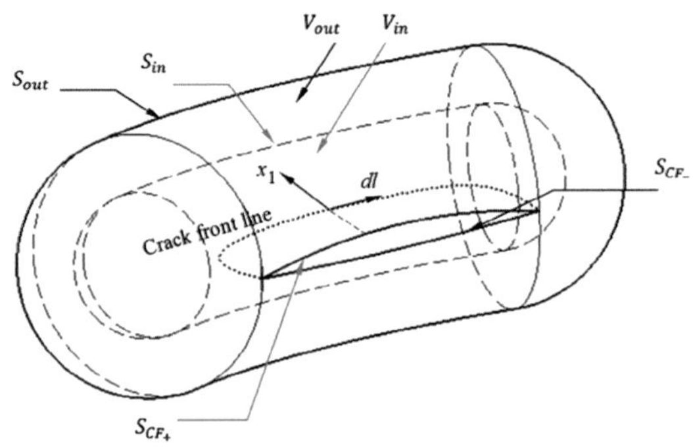

Fig. 1 θ field in 3D (left), Integration domain (right).

The particularity of this work lies in the implementation of the coupling between mechanical and moisture fields. The first term of the previous invariant integral, associated with stationary crack behavior, can be reformulated as follows:

$$\boldsymbol{A\theta^{3D}} = -\frac{1}{2}\int_V \left[{\sigma_{ij,k}}^{(v)}.u_i - {\sigma_{ij}}^{(u)} v_{i,k} - \gamma_{ij}\left(\delta w\left(v_{i,j} - \emptyset_{,j}\right) - \left(\delta w\right)_{,j}\left(v_{ij} - \emptyset\right)\right)\right].\theta_{k,j}.dV \quad (2)$$

In this new formulation, γ_{ij} is the tensor responsible for coupling the material's mechanical behavior with its hygroscopic effects. The components of this tensor are derived from the Navier equation in continuum mechanics and are presented in this paper in the form of Equation 3.

$$\gamma_{ij} = \sum_{kl} C_{ijkl}.\alpha_{kl} \tag{3}$$

The compliance tensor C_{ijkl}, which characterizes the material state at a given time under a specific mechanical load, is reorganized to retain only the components relevant to the principal orthotropic directions (L, R, and T). The indices k and l are projected onto the directions of the diagonal matrix α_{kl}, which represents the tensor whose components correspond to the coefficients of the hygroscopic shrinkage-swelling model of wood, taken as the principal orthotropic reference. The quantity $\delta w = w - w_{ref}$ denotes the moisture content variation within the material relative to a reference state w_{ref}, while $(\delta w)_{,j}$ represents its gradient in the considered direction.

The virtual mechanical fields along the crack front, as well as the components of the matrix $\emptyset$, are derived from the formulations of Sih [11]. For a two-dimensional problem involving an elastic anisotropic material, the virtual displacement components v_1 and v_2 have been redefined to incorporate the behavioral changes of wood induced by the presence of an external moisture load. Their revised expressions are presented as follows:

$$v_1 = K_I\sqrt{\frac{2r}{\pi}}\Re\left[\frac{1}{\mu_1-\mu_2}\left(\mu_1 p_2\sqrt{cos\varphi+\mu_2 sin\varphi}-\mu_2 p_1\sqrt{cos\varphi+\mu_1 sin\varphi}\right)\right]$$
$$+K_{II}\sqrt{\frac{2r}{\pi}}\Re\left[\frac{1}{\mu_1-\mu_2}\left(p_2\sqrt{cos\varphi+\mu_2 sin\varphi}-p_1\sqrt{cos\varphi+\mu_1 sin\varphi}\right)\right] \tag{4}$$

$$v_2 = K_I\sqrt{\frac{2r}{\pi}}\Re\left[\frac{1}{\mu_1-\mu_2}\left(\mu_1 q_2\sqrt{cos\varphi+\mu_2 sin\varphi}-\mu_2 q_1\sqrt{cos\varphi+\mu_1 sin\varphi}\right)\right]$$
$$+K_{II}\sqrt{\frac{2r}{\pi}}\Re\left[\frac{1}{\mu_1-\mu_2}\left(q_2\sqrt{cos\varphi+\mu_2 sin\varphi}-q_1\sqrt{cos\varphi+\mu_1 sin\varphi}\right)\right] \tag{5}$$

where (r, φ) denotes the polar coordinates of a point P_n near the crack tip, $\Re(.)$ the real part of the complex quantity in parentheses. K_I and K_{II} are the stress intensity factor in mode I (opening crack) and mode II (shear crack), respectively. μ_1 and μ_2 are the roots of the characteristic equation, which depends on the coefficients of the modified compliance matrix. In the case of an elastic anisotropic material, this matrix takes the following general form:

$$\xi_{11}\mu^4 - 2\xi_{16}\mu^3 + (2\xi_{12}+\xi_{66})\mu^2 - 2\xi_{26}\mu + \xi_{22} = 0 \tag{6}$$

The components ξ_{ij} at a given humidity state w are expressed in terms of the components C_{ij} of the compliance matrix, taken as a reference state at humidity w_{ref}. The parameters p_i and q_i are defined as follows:

$$p_i = \xi_{11}\mu_i^2 + \xi_{12} - \xi_{16}\mu_i \text{ and } q_i = \xi_{12}\mu_i - \xi_{26} + \frac{\xi_{22}}{\mu_i} \tag{7}$$

Analysis

The computation of the stress intensity factor and the energy release rate implies a numerical implementation of the integral in Equation (2). In this study, finite element analyses are carried out using the Cast3M software. The mesh geometry (Figure 2a) corresponds to a MMCG (Mixed Mode Crack Growth) specimen, selected for its ability to separate mixed fracture modes while maintaining a significant stability zone [12]. An initial crack length of 25 mm is considered, and the calculations are performed under a plane stress configuration in pure opening mode. Figure 2b illustrates the deformed mesh after the application of both mechanical and moisture loading.

The material studied is Scots pine (*Pinus sylvestris*), with its mechanical properties at a reference humidity of 12% listed in Table 1. These properties are subsequently adjusted at each step of the finite element analysis using the Guitard formulas [13], allowing for the consideration of mechanical behavior variations due to humidity changes.

The finite element analysis must account for the stress induced by the external moisture loading, which need to be subtracted from the initial mechanical loading. This process is outlined in the algorithmic diagram shown in Figure 3.

In the absence of experimental data, the reliability of the invariant integral implementation is typically validated by ensuring its independence from the integration domain. This validation involves computing the invariant integral across

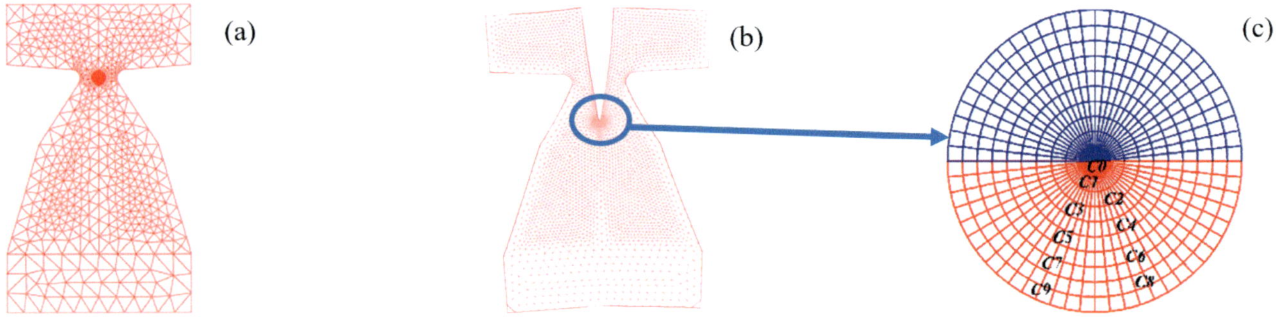

Fig. 2 (a) MMCG mesh, (b) Opening mode, (c) Radiating mesh at the crack tip.

Table 1 Properties of *Pinus sylvestris* at 12% of moisture content.

Properties	E_L (Mpa)	E_R (Mpa)	G_{RL} (Mpa)	ν_{RL}	α_L	α_R
Value	15000	600	700	0.5	0.0012	0.037

multiple rings within the integration domain. Here, the domain remains continuous through the field θ, provided that all calculations stay within its bounds neither too close to the crack tip nor too far toward the outer edge. In this study, the integration domain around the crack tip is divided into 10 rings, as illustrated in Figure 2c. Figures 4a and 4b present the calculation of the energy release rate for rings 2 to 8 under different moisture content levels.

First, a singularity is observed in the second ring of the integration domain. This is a well-known issue, as the calculations rely on stress fields that are too close to the singular region, where interpolations using asymptotic field approximations are performed with a limited number of finite elements, reducing interpolation accuracy. Beyond this, the independence of the integral is verified for rings 3 to 7. However, a disturbance reappears from ring 8 onward, as it is near the boundary of the integration domain, making the integration sensitive to boundary conditions. The fluctuation in independence ranges from

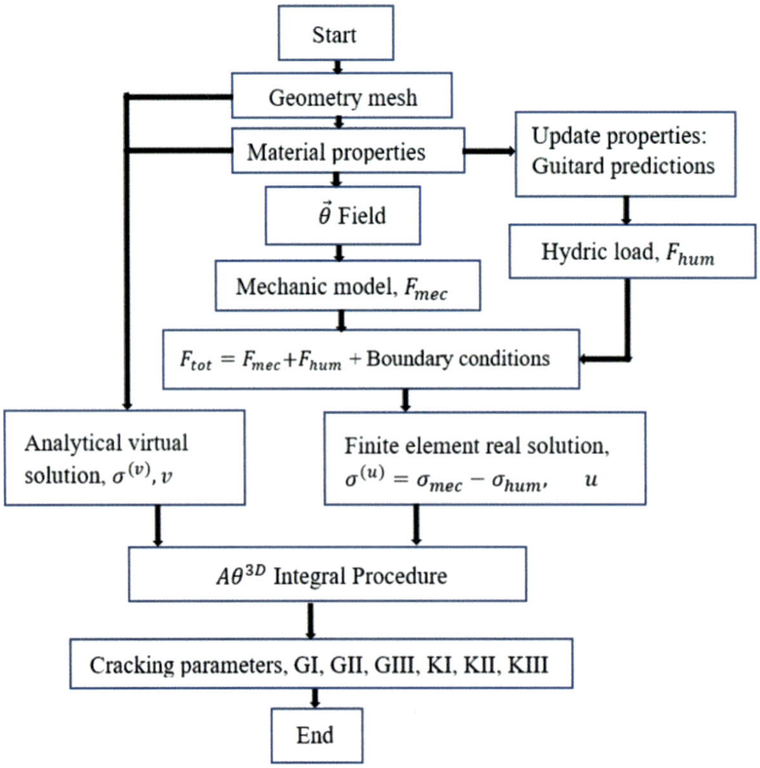

Fig. 3 Finite element numerical scheme implementation.

2% to 5%, compared to only 1% for the invariant integral $M\theta$ (equivalent to $A\theta$ in the absence of moisture loading). This suggests that the invariant integral is slightly affected when an external load is applied. Similar findings have been reported in previous studies on the independence of this integral under thermal loading [1, 9].

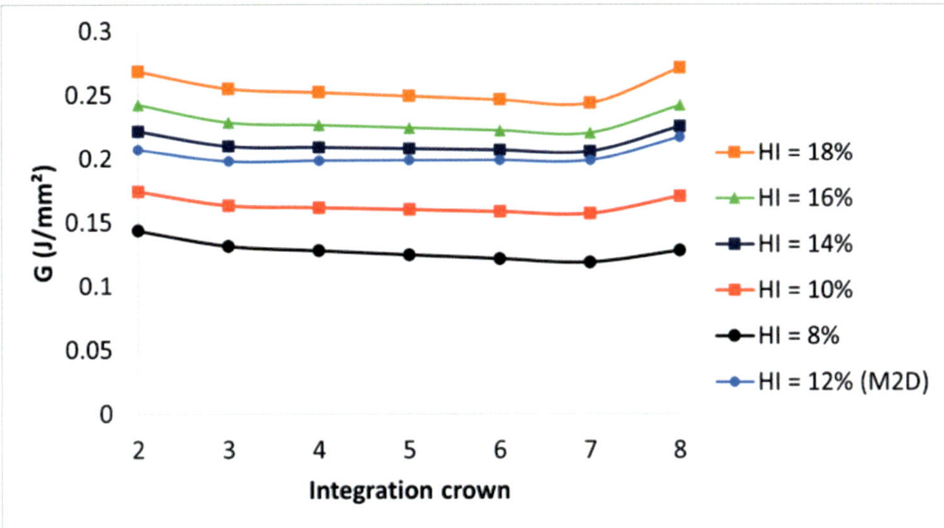

Fig. 4a Path independence verification of energy release rate for 8% to 18% moisture content.

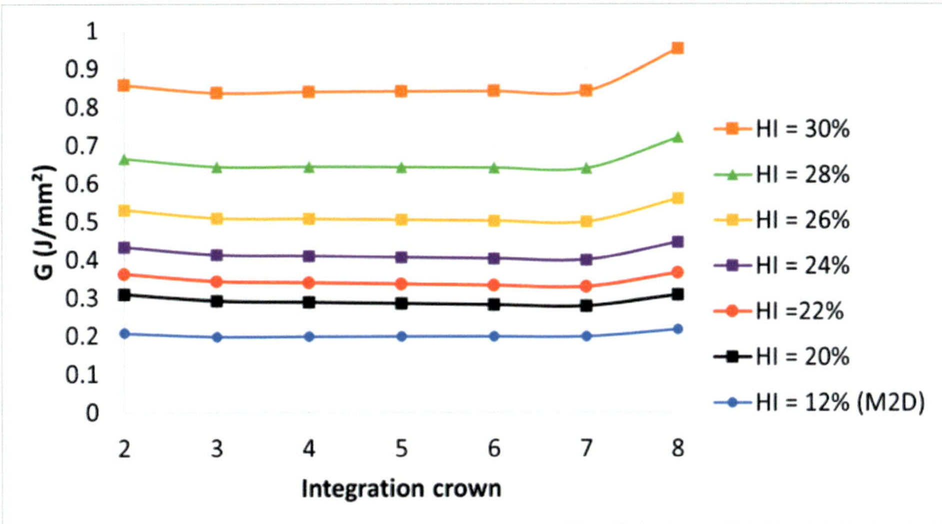

Fig. 4b Path independence verification of energy release rate for 20% to 30% moisture content.

Conclusion

This work presents a new analytical formulation of the invariant integral **T** to investigate crack propagation in wood under external hydric loading. The approach is based on Noether's theorem, with mechanical fields reformulated to incorporate hydric parameters and extended to a three-dimensional framework. As an initial validation, the first term of the integral is numerically implemented in mode I fracture for a two-dimensional MMCG wood specimen. The crack tip mesh is divided into ten concentric surface rings within the integration domain, enabling an assessment of the integration contour's independence. The computed energy release rate for each ring confirms this independence for rings 3 to 7, within a humidity range of 8% to 30%. However, numerical disturbances arise near the crack tip (rings 1 and 2) and at the outer boundary of the integration domain. Since these results assume a uniform humidity distribution within the material, future studies will examine the effects of moisture transport, mixed-mode fracture, and crack lips submitted to pressure.

Acknowledgments The authors would like to express their appreciation to the National Research Agency (ANR) for its financial support of this work through DIAMWOOD Project ANR PRCE N° 23-CE22-0006-03.

References

1. H. Riahi, R. Moutou Pitti, F. Dubois, A. Chateauneuf, « Mixed-mode fracture analysis combining mechanical, thermal and hydrological effects in an isotropic and orthotropic material by means of invariant integrals », *Theoretical and Applied Fracture Mechanics*, pp 424–434, Vol 85, 2016.
2. R. Moutou Pitti, « Découplage des modes mixtes de rupture dans les matériaux viscoélastiques orthotropes : modélisation et expérimentation » *PhD thesis*, University of Limoges, pp 162, 2008.
3. R. Moutou Pitti, F. Dubois, N. Sauvat, C. Petit, « Mixed mode fracture separation in viscoelastic orthotropic media: numerical and analytical approach by the mv-integral », *International Journal of Fracture*, pp 181–193, Vol 125, 2007.
4. R. Moutou Pitti, F. Dubois, C. Petit, N. Sauvat, O. Pop, « A new integral parameter for mixed-mode crack growth in viscoelastic orthotropic media », *Engineering Fracture Mechanics*, pp 4450–4465, Vol 75, 2008.
5. J. R. Rice, « A path independent integral and the approximate analysis of strain concentrations by notches and cracks », *Journal of Applied Mechanics*, pp 379–385, Vol 35, 1968.
6. J. Hou, C. Zhang, Q. Li, « The concept and numerical evaluation of M-integral based on domain integral method in cracked viscoelastic materials », *Mechanics of Materials*, pp 103–363, Vol 145, 2020.
7. J. S. Afoutou, F. Dubois, « Three-dimensional fracture modes decoupling by means of invariant integrals in wood under mixed mode loading », *Theoretical and Applied Fracture Mechanics*, pp104-708, Vol 134, 2024.
8. R. Moutou Pitti, F. Dubois, C. Petit, « Generalization of the T and A integrals to viscoelasticity », *Comptes Rendus Mecanique*, pp 545–551, Vol 336, 2008.
9. S. E. Hamdi, R. Moutou Pitti, F. Dubois, « Temperature variation effect on crack growth in orthotropic medium: Finite element formulation for the viscoelastic behavior in thermal cracked wood-based materials », *International Journal of Solids and Structures*, pp1–13, Vol 115, 2017.
10. E. Noether, « Invariant Variation Problems », *Transport Theory and Statistical Physics*, pp 186–207, Vol 1, 1971.
11. G. C. Sih, E. P. Chen, « Crack in Composite Materials », Vol. 6, Springer, 1981.
12. R. Moutou Pitti, F. Dubois, O. Pop, « A proposed mixed-mode fracture specimen for wood under creep loadings », *International Journal of Fracture*, pp 195–209, Vol 167, 2011.
13. D. Guitard, F. El Amri, « Modèles prévisionnels de comportement élastique tridimensionnel pour les bois feuillus et les bois résineux », *Annales Des Sciences Forestières*, pp 335–358, Vol *44*, 1987.

For Product Safety Concerns and Information please contact our
EU representative GPSR@taylorandfrancis.com Taylor & Francis
Verlag GmbH, Kaufingerstraße 24, 80331 München, Germany